AQA Science

Exclusively endorsed and approved by AQA

Revision Guide

Pauline Anning

Series Editor: Lawrie Ryan

GCSE Physics

Nelson Thornes

a Wolters Kluwer business

D0255892

Text © Pauline Anning 2006
Original illustrations © Nelson Thornes Ltd 2006

The right of Pauline Anning to be identified as author of this work has been asserted by her in accordance with the Copyright, Designs and Patents Act 1988.

All rights reserved. No part of this publication may be reproduced or transmitted in any form or by any means, electronic or mechanical, including photocopy, recording or any information storage and retrieval system, without permission in writing from the publisher or under licence from the Copyright Licensing Agency Limited, of 90 Tottenham Court Road, London W1T 4LP.

Any person who commits any unauthorised act in relation to this publication may be liable to criminal prosecution and civil claims for damages.

Published in 2006 by:
Nelson Thornes Ltd
Delta Place
27 Bath Road
CHELTENHAM
GL53 7TH
United Kingdom

06 07 08 09 10 / 10 9 8 7 6 5 4 3 2

A catalogue record for this book is available from the British Library

ISBN 0 7487 8313 X

Cover photographs: wave by Corel 391 (NT); static electricity by Photodisc 29 (NT); astronaut by Photodisc 34 (NT)
Cover bubble illustration by Andy Parker

Illustrations by Bede Illustration
Page make-up by Wearset Ltd

Printed and bound in Croatia by Zrinski

Acknowledgements

Alamy/Popperfoto 37br, 47bl; **Corel 637 (NT)** 53tr; **Corel 640 (NT)** 1tr, 7m; **Digital Vision 6 (NT)** 80m; **Digital Vision 9 (NT)** 21b, 33tr; **Jim Breithaupt** 1mr, 12r, 64bl; **Photodisc 67 (NT)** 11m; **Rover Group** 56m; **Science Photo Library/Andrew Lambert Photography** 53m, 63m, 73br, /Keith Kent 37tl, /Martyn F. Chillmaid 10br, /Maximilian Stock LTD 8m, /NASA/ESA/STScl 30b, /NOAO/AURA/NSF 100b, /NRAO/AUI/NSF 102m, /TRL LTD. 37mr; **Transport & Road Research Laboratory** 49tr

Many thanks for the contributions made by Paul Lister and Jim Breithaupt.

Picture research by Stuart Sweatmore, Science Photo Library and johnbailey@ntlworld.com.

Every effort has been made to trace all the copyright holders, but if any have been overlooked the publisher will be pleased to make the necessary arrangements at the first opportunity.

How to answer questions

Question speak

Command word or phrase	What am I being asked to do?
compare	State the similarities and the differences between two or more things.
complete	Write words or numbers in the gaps provided.
describe	Use words and/or diagrams to say how something looks or how something happens.
describe, as fully as you can	There will be more than one mark for the question so make sure you write the answer in detail.
draw	Make a drawing to show the important features of something.
draw a bar chart / graph	Use given data to draw a bar chart or plot a graph. For a graph, draw a line of best fit.
explain	Apply reasoning to account for the way something is or why something has happened. It is not enough to list reasons without discussing their relevance.
give / name / state	This only needs a short answer without explanation.
list	Write the information asked for in the form of a list.
predict	Say what you think will happen based on your knowledge and using information you may be given.
sketch	A sketch requires less detail than a drawing but should be clear and concise. A sketch graph does not have to be drawn to scale but it should be the appropriate shape and have labelled axes.
suggest	There may be a variety of acceptable answers rather than one single answer. You may need to apply scientific knowledge and/or principles in an unfamiliar context.
use the information	Your answer **must** be based on information given in some form within the question.
what is meant by	You need to give a definition. You may also need to add some relevant comments.

Diagrams

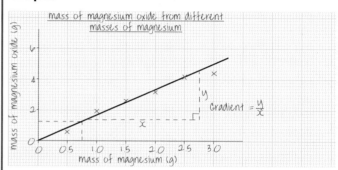

Things to remember:

- Draw diagrams in pencil.
- The diagram needs to be large enough to see any important details.
- Light colouring could be used to improve clarity.
- The diagram should be fully labelled.
- Label lines should be thin and end at the point on the diagram that corresponds to the label.

How long should my answer be?

Things to consider:

1 How many lines have been given for the answer?
- One line suggests a single word or sentence. Several lines suggest a longer and more detailed answer is needed.

2 How many marks is the answer worth?
- There is usually one mark for each valid point. So for example, to get all of the marks available for a three mark question you will have to make three different, valid points.

3 As well as lines, is there also a blank space?
- Does the question require you to draw a diagram as part of your answer?
- You may have the option to draw a diagram as part of your answer.

Graphs

Things to remember:

- Choose sensible scales so the graph takes up most of the grid.
- Don't choose scales that will leave small squares equal to 3 as it is difficult to plot values with sufficient accuracy on such scales.
- Label the axes including units.
- Plot all points accurately by drawing small crosses using a fine pencil.
- Don't try to draw a line through every point. Draw a line of best fit.
- A line of best fit does not have to go through the origin.
- When drawing a line of best fit, don't include any points which obviously don't fit the pattern.
- The graph should have a title stating what it is.
- To find a corresponding value on the y-axis, draw a vertical line from the x-axis to the line on the graph, and a horizontal line across to the y-axis. Find a corresponding value on the x-axis in a similar way.
- The gradient or slope of a line on a graph is the amount it changes on the y-axis divided by the amount it changes on the x-axis. (See the graph above.)

Calculations

- Write down the equation you are going to use, if it is not already given.
- If you need to, rearrange the equation.
- Make sure that the quantities you put into the equation are in the right units. For example you may need to change centimetres to metres or grams to kilograms.
- Show the stages in your working. Even if your answer is wrong you can still gain method marks.
- If you have used a calculator make sure that your answer makes sense. Try doing the calculation in your head with rounded numbers.
- Give a unit with your final answer, if one is not already given.
- Be neat. Write numbers clearly. If the examiner cannot read what you have written your work will not gain credit. It may help to write a few words to explain what you have done.

How to use the 'How Science Works' snake

The snake brings together all of those ideas that you have learned about 'How Science Works'. You can join the snake at different places – an investigation might start an observation, testing might start at trial run.

How do you think you could use the snake on how the resistance of a wire varies with length? Try working through the snake using this example – then try it on other work you've carried out in class.

Remember there really is no end to the snake – when you reach the tail it is time for fresh observations. Science always builds on itself – theories are constantly improving.

OBSERVATION

I wonder why…

HYPOTHESIS

Perhaps it's because…

PREDICTION

I think that if…

I should be honest and tell it as it is. Does the data support or go against my hypothesis?

Is it a linear (straight line) relationship – positive, negative or directly proportional (starting at the origin)? or is it a curve – complex or predictable?

Which of these should I use?
- Bar chart
- Line graph
- Scatter graph

RELATIONSHIP SHOWN BY DATA

PRESENTING DATA

Am I going further than the data allows me?

Are the links I have found – causal, by association or simply by chance?

CONCLUSION

Have I given a balanced account of the results?

My conclusion would be more reliable and valid if I could find some other research to back up my results.

Just how reliable (trustworthy) was the data? Would it be more reliable if somebody else repeated the investigation? Was the data valid – did it answer the original question?

EVALUATION

USE SECONDARY DATA

There are still many questions that we cannot answer in scie

This is a flow-chart style diagram connecting the stages of a scientific investigation. The main stages (along the dark path) are:

DESIGN → CONTROL VARIABLE → TRIAL RUN → PREPARE A TABLE FOR THE RESULTS → CARRY OUT PROCEDURE → TECHNOLOGICAL DEVELOPMENTS

DESIGN

Should the variables I use be continuous (any value possible), discrete (whole number values), ordered (described in sequence) or categoric (described by words)?

Can I use my prediction to decide on the variable I am going to change (independent) and the one I am going to measure (dependent)?

CONTROL VARIABLE

I will try to keep all other variables constant, so that it is a fair test. That will help to make it valid.

TRIAL RUN

This will help to decide the:
- Values of the variables
- Number of repeats
- Range and interval for the variables

Are my instruments sensitive enough?

Will the method give me accuracy (i.e. data near the true value)? Will my method give me enough precision and reliability (i.e. data with consistent repeat readings)?

PREPARE A TABLE FOR THE RESULTS

I'll try to keep random errors to a minimum or my results will not be precise. I must be careful!

Are there any systematic errors? Are my results consistently high or low?

CARRY OUT PROCEDURE

Are there any anomalies (data that doesn't follow the pattern)? If so they must be checked to see if they are a possible new observation. If not, the reading must be repeated and discarded if necessary.

I should be careful with this information. This experimenter might have been biased – must check who they worked for; could there be any political reason for them not telling the whole truth? Are they well qualified to make their judgement? Has the experimenter's status influenced the information?

I should be concerned about the ethical, social, economic and environmental issues that might come from this research.

The final decisions should be made by individuals as part of society in general.

Could anyone exploit this scientific knowledge or technological development?

TECHNOLOGICAL DEVELOPMENTS

here are questions that science cannot answer at all – such as 'Should we…?' questions.

P1a | Energy and energy resources

Checklist

This spider diagram shows the topics in the unit. You can copy it out and add your notes and questions around it, or cross off each section when you feel confident you know it for your exams.

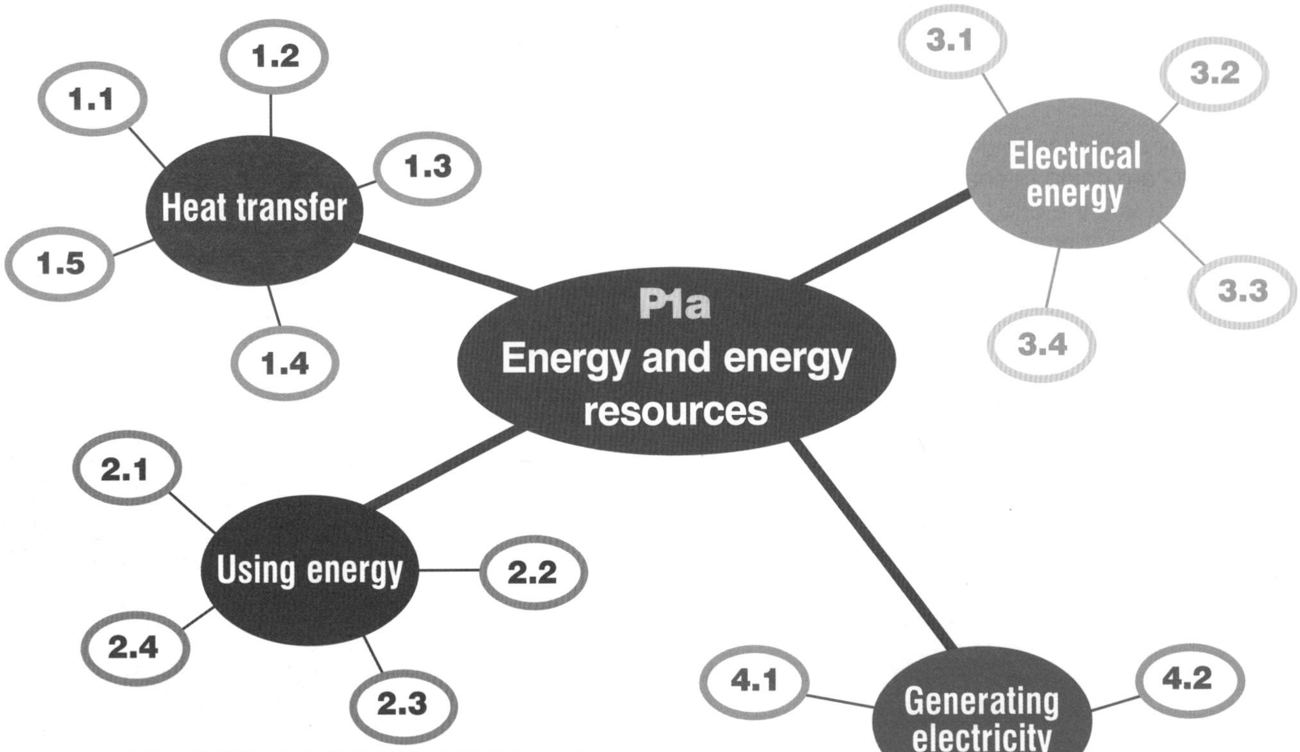

Tick when you:

reviewed it after your lesson	☑	☐	☐
revised once – some questions right	☑	☑	☐
revised twice – all questions right	☑	☑	☑

Move on to another topic when you have all three ticks.

Chapter 1 Heat transfer

1.1	Thermal radiation	☐	☐	☐
1.2	Surfaces and radiation	☐	☐	☐
1.3	Conduction	☐	☐	☐
1.4	Convection	☐	☐	☐
1.5	Heat transfer by design	☐	☐	☐

Chapter 2 Using energy

2.1	Forms of energy	☐	☐	☐
2.2	Conservation of energy	☐	☐	☐
2.3	Useful energy	☐	☐	☐
2.4	Energy and efficiency	☐	☐	☐

Chapter 3 Electrical energy

3.1	Electrical devices	☐	☐	☐
3.2	Electrical power	☐	☐	☐
3.3	Using electrical energy	☐	☐	☐
3.4	The National Grid	☐	☐	☐

Chapter 4 Generating electricity

4.1	Fuel for electricity	☐	☐	☐
4.2	Energy from wind and water	☐	☐	☐
4.3	Power from the Sun and the Earth	☐	☐	☐
4.4	Energy and the environment	☐	☐	☐

What are you expected to know?

Chapter 1 Heat transfer *See students' book pages 24–35*

- Heat energy is transferred by conduction, convection and radiation.
- Conduction and convection involve particles but radiation does not.
- Dark, matt surfaces are good absorbers and good emitters of heat radiation.
- Light, shiny surfaces are poor absorbers and poor emitters of heat radiation.
- The rate at which something radiates heat depends on its surface area and how much hotter it is than its surroundings.

Chapter 2 Using energy *See students' book pages 38–47*

- Energy cannot be created or destroyed, just transformed from one form to another.
- Not all energy transformations are useful, some energy is always 'wasted'.
- All energy is eventually transferred to the surroundings, which become warmer.
- Efficiency $= \dfrac{\text{useful energy transferred}}{\text{total energy supplied}}$

On a roller coaster – having fun with energy transformations!

Chapter 3 Electrical energy *See students' book pages 50–59*

- Power is the amount of energy transferred each second.
- Power is measured in watts. Their symbol is W.
- Energy is measured in joules. Their symbol is J.
- Electricity is transferred around the country through the National Grid.
- Energy transferred = power x time
- The kilowatt-hour is another unit of energy, which is used to measure the amount of electricity used at home.

An electricity meter

Chapter 4 Generating electricity *See students' book pages 62–71*

- Some power stations use non-renewable sources of energy such as coal and natural gas.
- In nuclear power stations, energy is produced by nuclear fission.
- Renewable energy sources include:
  - wind
  - waves
  - tides
  - falling water
  - the Sun
  - heat from the ground (geothermal energy).
- There are advantages and disadvantages to using both renewable and non-renewable energy sources.

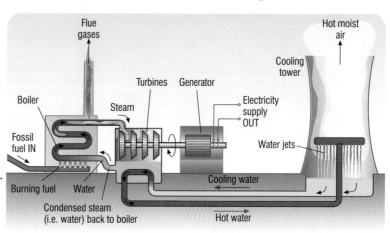

Inside a fossil fuel power station

Pre Test: Heat transfer

① Which type of heat transfer can occur through a vacuum?

② Which type of heat transfer does not involve particles?

③ Why are houses in hot countries often painted white?

④ Which teapot will keep tea hot the longest, a light, shiny one or a dull, dark one?

⑤ Why are metals the best conductors of heat?

⑥ What is an insulator?

⑦ Which type of heat transfer sets up currents in fluids?

⑧ What is convection?

⑨ What type of heat loss is reduced by cavity wall insulation?

⑩ What type of heat loss is reduced by carpets?

---

students' book page 24 ## P1a 1.1 Thermal radiation

KEY POINTS

1 Thermal radiation is energy transfer by electromagnetic waves.
2 All objects emit thermal radiation.
3 The hotter an object is, the more thermal radiation it emits.

Thermal or heat radiation is the transfer of energy by infra-red waves. These waves are part of the electromagnetic spectrum.

● All objects emit (give off) heat radiation.
● The hotter the object the more heat radiation it emits.
● Heat radiation can travel through a vacuum like space. This is how we get heat from the Sun.

Key words: thermal, heat, radiation, infra-red, emit

GET IT RIGHT!

Transfer of heat energy by infra-red radiation does **not** involve particles.

Detecting infra-red radiation

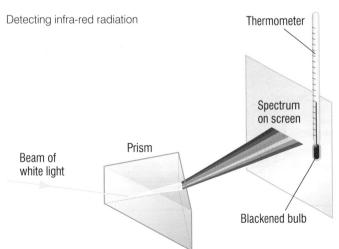

Thermometer

Spectrum on screen

Prism

Beam of white light

Blackened bulb

CHECK YOURSELF

1 Which part of the electromagnetic spectrum is concerned with heat energy?

2 How does heat from the Sun reach the Earth?

3 Do all objects give out the same amount of heat radiation? Explain your answer.

P1a 1.2 Surfaces and radiation

KEY POINTS

1 Dark, matt surfaces are better emitters of thermal radiation than light, shiny surfaces.
2 Dark, matt surfaces are better absorbers of thermal radiation than light, shiny surfaces.

- Dark, matt surfaces are good absorbers of radiation. An object painted dull black and left in the Sun will become hotter than the same object painted shiny white.
- Dark, matt surfaces are also good emitters of radiation. So an object that is painted dull black will lose heat and cool down quicker that the same object painted shiny white.

Key words: absorber, emitter

EXAMINER SAYS...

Exam questions often involve different applications of absorption and emission.

GET IT RIGHT!

Don't be fooled by most central heating radiators that are covered in glossy white paint. The best radiators of heat are dark, matt surfaces.

Remember that an object that is warmer than the surroundings will lose heat energy and cool down. On the other hand, an object that is cooler than the surroundings will gain heat energy and warm up.

CHECK YOURSELF

1 Which surfaces are the best emitters of heat radiation?

2 Which surfaces are the best absorbers of heat radiation?

3 Why are the pipes on the back of a fridge usually painted black?

P1a 1.3 Conduction

KEY POINTS

1 Conduction in a metal is due mainly to free electrons transferring energy inside the metal.
2 Non-metals are poor conductors because they do not contain free electrons.
3 Materials such as fibreglass are good insulators because they contain pockets of trapped air.

Conduction occurs mainly in solids. Most liquids and all gases are poor conductors.

- If one end of a solid is heated, the particles at that end gain kinetic energy and vibrate more. This energy is passed to neighbouring particles and in this way the heat is transferred through the solid.

This process occurs in metals.

- In addition, when metals are heated their free electrons gain kinetic energy and move through the metal transferring energy by colliding with other particles. Hence all metals are good conductors of heat.
- Poor conductors are called insulators.

Key words: conduction, conductor, insulator

CHECK YOURSELF

1 Why are materials that trap air, such as fibreglass, good insulators?

2 Why are saucepans often made of metal with wooden handles?

3 Which materials are the best conductors of heat?

EXAMINER SAYS...

Know some examples of insulators and how they are used.

YOUR GRADE

Make sure you can explain why all metals are good conductors of heat energy.

- ⊕ Ion
- ○ Electron
- ⬤ Atom

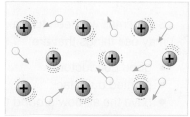
Energy transfer in a metal

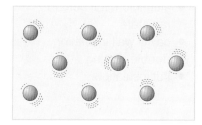
Energy transfer in a non-metal

P1a 1.4 Convection

KEY POINTS

1 Convection takes place only in liquids and gases (fluids).
2 Heating a liquid or a gas makes it less dense.
3 Convection is due to a hot liquid or gas rising.

AQA EXAMINER SAYS...

Remember that convection cannot occur in solids.

The type of surface makes no difference to the amount of conduction or convection from an object – it only affects radiated heat energy.

BUMP UP YOUR GRADE

Make sure that you can explain how convection currents are set up in terms of a fluid's changes in density.

Convection occurs in fluids.

When a fluid is heated it expands. The fluid becomes less dense and rises. The warm fluid is replaced by cooler, denser fluid. The resulting convection current transfers heat throughout the fluid.

Convection currents can be on a very small scale, such as heating water in a beaker, or on a very large scale such as heating the air above land and sea. Convection currents are responsible for onshore and offshore breezes.

Key words: fluid, convection, convection current

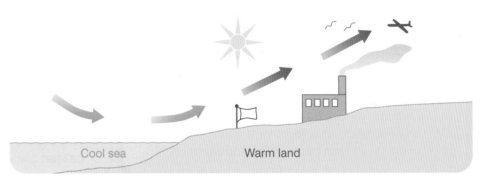

Cool sea   Warm land

CHECK YOURSELF

1 In which states of matter does convection occur?
2 What happens to the density of a fluid when it is heated?
3 Why doesn't convection take place in solids?

P1a 1.5 Heat transfer by design

KEY POINTS

1 A radiator has a large surface area so it can lose heat easily.
2 Small objects lose heat more easily than large objects.
3 Heat loss from a building can be reduced using:
   ● aluminium foil behind radiators
   ● cavity wall insulation
   ● double glazing
   ● loft insulation.

In many situations we want to minimise heat loss. We do this by reducing the losses due to conduction, convection and radiation.

● We can reduce heat loss by conduction by using insulators, e.g. trapping a layer of air.
● Heat loss by convection can be reduced by preventing convection currents being set up, e.g. by trapping air in small pockets.
● We can reduce heat loss by radiation by using light, shiny surfaces, which are poor emitters.

Sometimes we need to maximise heat loss to keep things cool. To do this we may use things that are:

● good conductors
● painted dull black
● have the air flow around them maximised.

Key words: maximise, minimise

EXAM HINTS

A vacuum flask is an application that often comes up in examination questions. Make sure you can relate the structure of a vacuum flask to conduction, convection and radiation. A vacuum flask keeps heat in so that hot things stay hot. It also keeps heat out so that cold things stay cold.

AQA EXAMINER SAYS...

To reduce heat loss in a particular situation, remember to include ways to minimise heat losses by conduction, convection and radiation.

BUMP UP YOUR GRADE

Know some examples of reducing heat losses from buildings and how they work. These should include loft insulation, cavity wall insulation, double glazing, draft excluders and carpets.

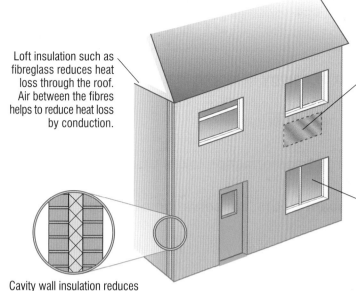

Loft insulation such as fibreglass reduces heat loss through the roof. Air between the fibres helps to reduce heat loss by conduction.

Aluminium foil between a radiator panel and the wall reflects heat radiation away from the wall.

A double glazed window has two glass panes with dry air or a vacuum between the panes. Dry air is a good insulator so it cuts down heat conduction. A vacuum cuts out heat transfer by convection as well.

Cavity wall insulation reduces heat loss through the walls. We place insulation between the two layers of brick that make up the walls of a house.

CHECK YOURSELF

1 What type of heat transfer is reduced by double glazing?

2 How does cavity wall insulation reduce heat loss from a building?

3 What is loft insulation usually made from?

P1a 1   End of chapter questions

1 Name the three types of heat transfer.

2 Which type of heat transfer occurs mainly in solids?

3 How are convection currents set up in fluids?

4 Describe the process of heat transfer through a metal.

5 What factors affect the amount of heat radiated by a body?

6 Why are gases poor conductors?

7 How does surface colour affect the rate of conduction?

8 Why do central heating radiators have large surface areas?

9 Why does a concrete floor feel colder to your feet than a carpeted floor at the same temperature?

10 Why are hot water tanks often wrapped in glass fibre jackets?

P1a 2 — Pre Test: Using energy

1. What form of energy is stored in any object that can fall?

2. What is kinetic energy?

3. What energy transformation takes place in a microphone?

4. What is meant by the 'conservation of energy'?

5. What is the useful energy transformation in an electric motor?

6. What happens to the energy wasted by a device?

7. What are the units of efficiency?

8. What is the SI unit of energy?

students' book page 38 P1a 2.1 — Forms of energy

KEY POINTS

1. Energy exists in different forms.
2. Energy can change (transform) from one form into another form.

BUMP UP YOUR GRADE

Make sure you are familiar with the different forms that energy can take and know some examples of each of them.

CHECK YOURSELF

1. What energy do we give to a spring if we squash it?

2. Where does the chemical energy stored in your muscles come from?

3. What form of energy does a moving train have?

Energy exists in different forms such as: light, thermal (heat), sound, kinetic (movement), nuclear, electrical, gravitational potential, elastic potential and chemical.

The last three are forms of stored energy.

Form of energy	Example
Light	From the Sun or a lamp
Thermal (heat)	Flows from a hot object to a colder object
Sound	From a loudspeaker or your voice
Kinetic (movement)	Anything moving
Nuclear	From nuclear reactions
Electrical	Whenever an electric current flows
Gravitational potential	Stored in any object that can fall
Elastic potential (strain)	Stored in stretched objects such as elastic bands or springs
Chemical	Stored in fuels, food and batteries and released when chemical reactions taken place

Energy can transform (change) from one form into another.

Key words: energy, kinetic, potential

P1a 2.2 Conservation of energy

KEY POINTS

1 Energy can be transformed from one form to another or transferred from one place to another.
2 Energy cannot be created or destroyed.

GET IT RIGHT!

Remember that the conservation of energy applies in any situation.

BUMP UP YOUR GRADE

Know some examples of energy transformations.

EXAM HINTS

The conservation of energy is an extremely important idea in physics so it often comes up in examination questions.

It is not possible to create or destroy energy. It is only possible to transform it from one form to another, or transfer (move) it from one place to another.

This means that the total amount of energy is always the same. This is called the 'conservation of energy'.

For example when an object falls, gravitational potential energy is transformed into kinetic energy. Similarly, stretching an elastic band transforms chemical energy into elastic potential energy. In a solar cell, light energy is transformed into electrical energy.

Key words: transform, transfer, conservation

On a roller coaster – having fun with energy transformations!

CHECK YOURSELF

1 What energy transformation takes place when you burn a fuel?

2 What energy transformations take place in a light bulb?

3 When you run, what type of energy is changed into kinetic energy? Where does this energy come from?

P1a 2.3 Useful energy

KEY POINTS

1 Useful energy is energy in the place we want it and in the form we need it.
2 Wasted energy is energy that is not useful energy.
3 Useful energy and wasted energy both end up being transferred to the surroundings, which become warmer.
4 As energy spreads out, it gets more and more difficult to use for further energy transfers.

AQA EXAMINER SAYS...

Remember that energy cannot be destroyed, so it is better to talk about energy that is 'wasted' than to say that energy is 'lost'.

A device (or machine) is something that transfers energy from one place to another or transforms energy from one form to another.

The energy supplied to the device is called the 'input energy'.

The energy we get out of the device consists of:

● useful energy, which is transferred to the place we want and in the form we want it
● wasted energy which is not usefully transferred or transformed (mostly it is converted as heat, frequently as a result of friction between the moving parts of the device).

From the conservation of energy we know that:

Input energy = useful energy + wasted energy

Both the useful energy and the wasted energy will eventually be transferred to the surroundings, and make them warm up. As it does so, it becomes more difficult to use the energy.

Key words: device, input energy, useful energy, wasted energy

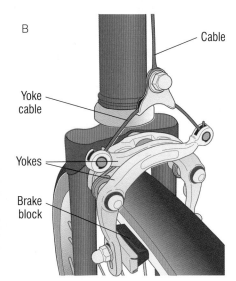

Cable

Yoke cable

Yokes

Brake block

Friction in action. A) Using a drill, B) braking.

GET IT RIGHT!

In some devices some of the wasted energy may be transferred into sound, but the amount of energy is usually very small. Eventually this energy will end up as heat and make the surroundings warmer.

CHECK YOURSELF

1 What is the useful energy transformation in a light bulb and what happens to the wasted energy?

2 Why does energy become more difficult to use as it spreads out?

3 Why do computers and televisions have vents?

KEY POINTS

1 Energy is measured in joules.
2 The efficiency of a device = useful energy transferred by the device ÷ total energy supplied to the device
3 Wasted energy causes inefficiency.

AQA EXAMINER SAYS...

Remember that no device (except an electric heater) can be 100% efficient. So if you do an efficiency calculation and the answer is greater than 1 or 100%, you have made an error and should check your working.

GET IT RIGHT!

Efficiency is a ratio. That means it does not have a unit.

The unit of energy is the joule, symbol J. This unit is used for all forms of energy.

The less energy that is wasted by a device, the more efficient the device is said to be.

We can calculate the efficiency of a device using the equation in the key points. It can also be written as:

$$\text{Efficiency} = \frac{\text{energy usefully transformed by the device}}{\text{input energy}}$$

The efficiency can be left as a fraction or multiplied by 100 to give a percentage.

No device can be 100% efficient, except an electric heater, which usefully transforms all of the electrical energy supplied to it into heat energy.

Key words: joule, efficiency

EXAM HINTS

Be careful in questions when you substitute energies into the efficiency equation. You may be given the energy usefully transferred, or you may be given the energy wasted and have to subtract this from the input energy to find the energy usefully transferred.

CHECK YOURSELF

1 If a device is adjusted so that it wastes less energy as heat, what happens to its efficiency?

2 In a light bulb, for every 20 joules of energy input to the bulb, 5 joules are usefully transformed into light energy. What is the efficiency of the bulb?

3 In an electric motor 2000 joules of energy are given out as heat to the surroundings for every 5000 joules of electrical energy supplied to the motor. What is the efficiency of the motor?

P1a 2   End of chapter questions

1 List three forms of stored energy and give an example of each.

2 Explain the energy transformations that take place if you climb to the top of a ladder then fall to the ground.

3 Why may an electric heater be 100% efficient when no other device ever is?

4 Calculate the efficiency of a kettle if it takes 720 000 J of energy to raise the temperature of a kettle full of cold water to boiling point when 750 000 J of energy is supplied to the kettle.

5 What happens to the energy that is not used to heat the water?

6 What energy do we give to a spring if we stretch it?

7 What is the useful energy transformation in a vacuum cleaner?

8 What energy transformation takes place in a solar cell?

① What energy transformation takes place in an electric iron?

② What energy transformations take place in a television?

③ What is meant by the power of a device?

④ What is the unit of power?

⑤ What unit do electricity companies use to measure the amount of electrical energy used?

⑥ What is the equation that relates energy, power and time?

⑦ What is the National Grid?

⑧ What type of transformer makes voltages larger?

---

students' book page 50

P1a 3.1 — Electrical devices

KEY POINTS

1 Electrical energy is energy transfer due to an electric current.
2 Uses of electrical devices include:
   - heating
   - lighting
   - making objects move (using an electric motor)
   - creating sound and visual images.

Electrical devices are extremely useful. They transform electrical energy into whatever form of energy we need at the flick of a switch.

Common electrical devices include:

- kettles, to produce heat energy
- lamps, to produce light
- electric mixers, to produce kinetic energy
- speakers, to produce sound energy
- televisions, to produce light and sound energy.

Key words: electrical devices

Electrical devices

GET IT RIGHT!

Remember that all electrical devices will transform some electrical energy to heat, but this may not be a useful energy transformation.

CHECK YOURSELF

1 What energy transformations take place in an electric drill?

2 How is electrical energy supplied to a torch?

3 What energy transformations take place in a vacuum cleaner?

Electrical power

KEY POINTS

1 The unit of power is the watt (W), equal to 1 J /s.
2 1 kilowatt (kW) = 1000 watts
3 Power (in watts) =

$$\frac{\text{energy transferred (in joules)}}{\text{time taken (in seconds)}}$$

AQA EXAMINER SAYS…

Make sure that you can convert between kilowatts and watts, and between kilojoules and joules.

GET IT RIGHT!

The more powerful a device, the greater the rate at which it transfers or transforms energy.

The power of a device is the rate at which it transforms energy.

The unit of power is the watt, symbol W. A device with a power of 1 watt transforms 1 joule of electrical energy to other forms of energy every second.

Often a watt is too small a unit to be useful, so power may be given in kilowatts (kW). 1 kilowatt = 1000 watts.

Power is calculated using the equation:

$$\text{Power (in watts)} = \frac{\text{energy transferred (in joules)}}{\text{time taken (in seconds)}}$$

Key words: power, watt, kilowatt

Muscle power

CHECK YOURSELF

1 How many watts are in 30 kilowatts?

2 Which is more powerful, a 2.5 kW heater or a 3000 W heater?

3 An electric motor transforms 36 kJ of electrical energy into kinetic energy in 3 minutes. What is the useful power output of the motor?

Using electrical energy

1 Energy transferred (kW h)
= power of device (kW) × time (h)
2 Total cost of electricity =
number of kW h × cost per kWh

AQA EXAMINER SAYS...

Take care with the units here they
are tricky! Remember that the
kilowatt-hour is a unit of energy
(not power).

Companies that supply mains electricity charge customers for the amount of
electrical energy used. The amount of energy is measured in kilowatt-hours,
symbol kWh. A kilowatt-hour is the amount of energy that is transferred by a one
kilowatt device when used for one hour.

The amount of energy, in kWh, transferred to a mains device can be found by
multiplying the power of the device in kilowatts by the time it is used for in hours
(see the equation in the key points).

The electricity meter in a house records the number of kW h of energy used.
If the previous meter reading is subtracted from the current reading, the
electrical energy used between the readings can be calculated. The cost of the
electrical energy supplied is found by multiplying the number of kWh by the
cost per kWh, which is given on the electricity bill.

Key words: kilowatt-hour, kilowatt, hour

An electricity meter

CHECK YOURSELF

1 What quantity is measured in kilowatts?

2 How much electrical energy, in kWh, does it take to use a 9 kW shower
for 20 minutes?

3 The price of one kilowatt-hour of electricity is 8 p. How much does it cost
to use a 100 W electric light for 4 hours?

The National Grid

1 The National Grid is a network
of cables and transformers.
2 We use step-up transformers to
step power stations voltages up
to the grid voltage.
3 We use step-down transformers
to step the grid voltage down
for use in our homes.
4 A high grid voltage reduces
energy loss and makes the
system more efficient.

In Britain, electricity is distributed
through the National Grid. This is a
network of pylons and cables that
connects power stations to homes and
other buildings. Since the whole
country is connected to the system,
power stations can be switched in or
out according to demand.

BUMP UP YOUR GRADE

Be able to describe the reasons
why voltages are increased and
decreased within the National Grid
system.

In power stations, electricity is generated at a particular voltage. The voltage is
increased by step-up transformers before the electricity is transmitted across the
National Grid. This is because transmission at high voltage reduces energy
losses in the cables, making the system more efficient.

It would be dangerous to supply electricity to consumers at these very
high voltages. So step-down transformers are used to reduce the voltage to
230 volts.

Key words: National Grid, step-up transformer, step-down transformer

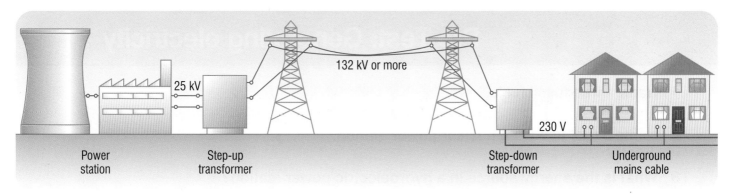

25 kV

132 kV or more

230 V

Power
station

Step-up
transformer

Step-down
transformer

Underground
mains cable

The National Grid

EXAM HINTS

Know the different parts of the
National Grid system and the
order in which they are used. In an
examination question you might
have to describe them in words or
label them on a diagram.

CHECK YOURSELF

1 What type of transformer makes voltages smaller?

2 Why is electricity transmitted at very high voltages across the National Grid?

3 What type of transformer is found in a local sub-station?

P1a 3   End of chapter questions

1 State an equivalent unit to the watt.

2 What device transforms sound energy to electrical energy?

3 What device transforms electrical energy to sound energy?

4 In a 3 kW kettle how many joules of electrical energy are transformed from electrical energy to heat energy each second?

5 An immersion heater converts 36 000 000 J of electrical energy into heat energy when it is switched on for 1 hour.
What is the power of the heater in kilowatts?

6 How much does it cost to use a 1200 W vacuum cleaner for 10 minutes if electrical energy costs 7 p per kilowatt-hour?

7 Draw a block diagram showing the different parts of the National Grid system and the order in which they are used.

8 Why are voltages reduced to 230 V before reaching homes?

① **Name three fossil fuels.**

② **Name a fuel used in a nuclear power station.**

③ **What is the energy source in a hydroelectric power station?**

④ **What is a wind turbine?**

⑤ **What does a solar cell do?**

⑥ **What is 'geothermal energy'?**

⑦ **Name two renewable energy resources.**

⑧ **Suggest an advantage of a nuclear power station.**

---

students' book page 62

P1a 4.1   Fuel for electricity

KEY POINTS

1 Electricity generators in power stations are driven by turbines.
2 Much more energy is released per kilogram from uranium than from fossil fuels.

GET IT RIGHT!

Most power stations burn fuels to produce heat. In a nuclear power station uranium is not burned, the heat comes from a process called nuclear fission.

CHECK YOURSELF

1 Which of the following is *not* a fossil fuel: coal, plutonium, gas, oil?

2 What process produces heat in a nuclear power station?

3 Why is water turned into steam in a coal-fired power station?

In most power stations, water is heated to produce steam. The steam drives a turbine, which is coupled to an electrical generator that produces the electricity.

The heat can come from burning a fuel such as coal, oil or gas (called fossil fuels); or hot gases may drive the turbine directly.

In a nuclear power station, the fuel used is uranium (or sometimes plutonium). The nucleus of a uranium atom can undergo a process called 'fission' that releases energy. There are lots of uranium nuclei, so lots of fission reactions take place, producing lots of heat energy. This energy is used to turn water into steam.

More energy is released from each kilogram of uranium undergoing fission reactions than from each kilogram of fossil fuel that we burn.

Key words: turbine, generator, fossil fuel, fission

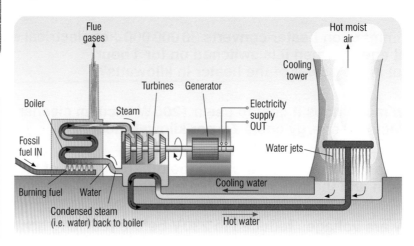

Inside a fossil fuel power station

P1a 4.2 Energy from wind and water

KEY POINTS

1 A wind turbine is an electricity generator on top of a tall tower.
2 A wave generator is a floating generator turned by the waves.
3 Hydroelectricity generators are turned by water running downhill.
4 A tidal power station traps each high tide and uses it to turn generators.

AQA EXAMINER SAYS...

There are a number of different ways that electricity can be generated using energy from water. Exam questions could be asked about any of them so learn them all carefully and make sure you understand the differences between them.

We can use energy from wind and water to drive turbines directly.

● In a wind turbine, the wind passing over the blades makes them rotate and drive a generator at the top of a narrow tower.

Electricity can be produced from energy obtained from falling water, waves or tides.

● At a hydroelectric power station, water which has been collected in a reservoir is allowed to flow downhill and turn turbines at the bottom of the hill.
● In a pumped storage system, surplus electricity is used, at times of low demand, to pump the water back up the hill to the top reservoir.
● This means that the energy is stored, and then at times of high demand the water can be released to fall through the turbines and convert the stored energy to electrical energy.
● We can use the movement of the waves on the sea to generate electricity with devices that float on the water. The movement drives a turbine that turns a generator. Then the electricity is delivered to the grid system on shore by a cable.
● The level of the sea around the coastline rises and falls twice each day. These changes in sea level are called tides. If a barrage is built across a river estuary, the water at each high tide can be trapped behind it. When the water is released to fall down to the lower sea level, it drives turbines.

Key words: wind, waves, hydroelectricity, tides

CHECK YOURSELF

1 Why can't a hydroelectric power station be built in a flat area?

2 What form of energy is stored in the water in the top reservoir of a pumped storage scheme?

3 Why is wave power likely to be less reliable than tidal power?

P1a 4.3 Power from the Sun and the Earth

KEY POINTS

1 We can convert solar energy into electricity using solar cells or use it to heat water directly in solar heating panels.
2 Geothermal energy comes from the energy released by radioactive substances deep inside the Earth.

BUMP UP YOUR GRADE

Adding the details in a description of geothermal energy will earn you extra marks.

Energy from the Sun travels through space to the Earth as electromagnetic radiation. A solar cell can convert this energy into electrical energy. Each cell only produces a small amount of electricity, so they are useful to power small devices such as watches and calculators. We can also join together large numbers of the cells to form a solar panel.

Water flowing through a solar heating panel is heated directly by energy from the Sun.

Heat energy is produced inside the Earth by radioactive processes and this heats the surrounding rock. In a few parts of the world, hot water comes up to the surface naturally and can be used to heat buildings nearby.

In other places, very deep holes are drilled and cold water is pumped down to the hot rocks where it is heated and comes back to the surface as steam. The steam is used to drive turbines that turn generators and so electricity is produced.

Key words: solar energy, geothermal energy

CHECK YOURSELF

1 Where does the geothermal energy that produces heat in the ground come from?

2 What energy transformation takes place in a solar cell?

3 Why are large numbers of solar cells often joined to make a solar panel?

P1a 4.4 Energy and the environment

KEY POINTS

1 Fossil fuels produce greenhouse gases.
2 Nuclear fuels produce radioactive waste.
3 Renewable energy resources can affect plant and animal life.

Coal, oil, gas and uranium are non-renewable energy resources. This means that the rate at which they are used is very much faster than the rate at which they are produced. If we continue to use them up at the current rate they will soon run out.

Renewable energy resources will not run out.

There are advantages and disadvantages to using each type of energy resource:

Energy resource	Advantages	Disadvantages
Coal	• Bigger reserves than the other fossil fuels • Reliable	• Non-renewable • Production of CO_2, a greenhouse gas • Production of SO_2, causing acid rain
Oil	• Reliable	• Non-renewable • Production of CO_2, a greenhouse gas • Production of SO_2, causing acid rain
Gas	• Reliable • Gas power stations can be started up quickly to deal with sudden demand	• Non-renewable • Production of CO_2, a greenhouse gas
Nuclear	• No production of polluting gases • Reliable	• Non-renewable • Produces nuclear waste, which is difficult to dispose of safely • Risk of a big accident, such as Chernobyl
Wind	• Renewable • Free • No production of polluting gases	• Requires many large turbines • Unsightly and noisy • Not reliable, the wind does not always blow
Falling water	• Renewable • Free • No production of polluting gases • Reliable in wet areas • Pumped storage systems allow storage of energy • Can be started up quickly to deal with sudden demand	• Only work in wet and hilly areas • Flooding of an area affects the local ecology
Waves	• Renewable • Free • No production of polluting gases	• Can be a hazard to boats • Not reliable
Tides	• Renewable • Free • No production of polluting gases • Reliable, always tides twice a day	• Only a few river estuaries are suitable • Building a barrage affects the local ecology

Energy resource	Advantages	Disadvantages
Solar	• Renewable • Free • No production of polluting gases • Reliable in hot countries, in the daytime	• Only suitable for small amounts of electricity, or requires large number of cells • Unreliable in less sunny countries
Geothermal	• Renewable • Free • No production of polluting gases	• Only economically viable in a very few places • Drilling through large depth of rock is difficult and expensive

Key words: renewable, non-renewable, polluting

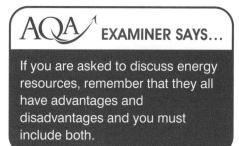

EXAMINER SAYS...

If you are asked to discuss energy resources, remember that they all have advantages and disadvantages and you must include both.

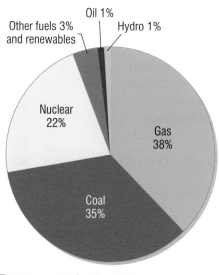

Energy sources for electricity

Other fuels 3% and renewables
Oil 1%
Hydro 1%
Nuclear 22%
Gas 38%
Coal 35%

CHECK YOURSELF

1 Are renewable or non-renewable energy resources the most reliable?

2 Which types of power station can be started up quickly to deal with periods of high demand?

3 What type of area would be most suitable for a wind farm?

P1a 4   End of chapter questions

1 Explain the difference between renewable and non-renewable energy resources.

2 In which three ways can water be used as an energy resource to generate electricity?

3 How can geothermal energy be used to produce heat and electricity?

4 What are the advantages of using fossil fuels in power stations to produce electricity?

5 What are the disadvantages of using fossil fuels in power stations to produce electricity?

6 Suggest a disadvantage of a nuclear power station.

7 Why is geothermal energy economically unviable in most places?

8 What colour are solar heating panels usually painted?

1 The drawing shows an array of solar cells.

Array of solar cells

(a) What is the energy transfer that takes place in the solar cells? (1 mark)

(b) When it is sunny the array can produce between 2 kW and 5 kW of electricity.
   (i) When the array is producing 2 kW of electricity how many Joules of energy does it produce each second. (2 marks)
   (ii) The total energy produced in one day was 30 kWh. What was the minimum time taken to produce this amount of energy? (3 marks)

(c) Suggest a suitable use for a single solar cell. (1 mark)

2 Energy transformations take place in a television set.

(a) What useful energy transformations take place in a television set? (3 marks)

(b) At the back of the television there are ventilation slots. Explain why the television has ventilation slots. (3 marks)

3 A student is investigating energy transformations in a simple pendulum. She releases it from position A.

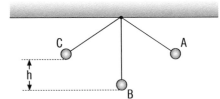

(a) Describe the energy transformations that take place as the pendulum swings from position A, through position B, to position C. (3 marks)

(b) The student wants to measure the vertical height, h, to which the pendulum swings. It is difficult to make this measurement. What could she do to make her measurement reliable? (2 marks)

4 (a) Describe the process of heat conduction in metals. (5 marks)

(b) Why are most fluids poor conductors of heat? (2 marks)

5 A homeowner wants to insulate her house to reduce heat losses.

Explain how fitting each of the following will reduce heat losses

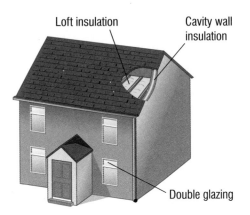

Loft insulation   Cavity wall insulation

Double glazing

(a) cavity wall insulation (3 marks)

(b) double glazing (2 marks)

(c) glass fibre loft insulation. (2 marks)

6 Tidal barrages and wind farms can be used to produce electricity from *renewable energy resources*.

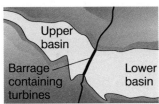

Upper basin
Barrage containing turbines   Lower basin

Tidal barrage

Wind turbines

(a) What is meant by *renewable energy resources*? (1 mark)

(b) These renewable energy resources do not produce polluting gases but they may still have a poor effect on the environment.
What poor effect can tidal barrages and wind farms have on the environment? (5 marks)

The answer is worth 7 marks out of the 9 available. The responses worth a mark are underlined in red.

We can improve the answers in several ways:

A further mark would be given for stating that the steam drives a **turbine which is coupled** to the generator.

In a nuclear power station heat is produced by nuclear fission.
(a) Explain how the heat is used to produce electricity. *(4 marks)*
(b) Apart from cost, what are the advantages and disadvantages of using a coal-fired power station to produce electricity? *(5 marks)*

a) The heat is used to boil water and turn it into steam. The steam turns a generator that makes electricity.
b) ADVANTAGES
Coal is a reliable energy source because it does not rely on things like the weather.
There are bigger reserves of coal than the other fossil fuels.
DISADVANTAGES
Coal is non-renewable.
Coal produces polluting gases.

Extra detail is needed to get the final mark, e.g. naming the polluting gases.

The answer is worth 5 marks out of the 11 available. The responses worth a mark are underlined in red.

We can improve the answer in several ways:

The question asks the student to 'describe'. This answer lacks detail, so does not get any further marks.

The fraction in the equation is upside down, so does not get any credit for working. It is a pity the student did not remember that the efficiency of a device cannot be greater than 1 (100%), and so spot the mistakes.

1 Electrical devices are very useful. They transform electrical energy to other forms at the flick of a switch
(a) A vacuum cleaner contains an electric motor.
   (i) What is the useful energy transformation in the motor?
   (ii) Describe what happens to the energy that is not usefully transformed. *(4 marks)*
(b) The motor usefully transforms 800 J of energy for every 1000 J of energy supplied to it.
   Calculate the efficiency of the motor. *(3 marks)*
(c) The power of the vacuum cleaner is 1 800 W.
   Calculate the cost of using it for 2 hours if electricity costs 3p per kilowatt-hour. *(4 marks)*

(a) (i) electrical to kinetic
    (ii) turns to heat

(b) efficiency = $\dfrac{\text{energy in}}{\text{energy out}}$

    efficiency = $\dfrac{1000}{800}$ = 1.25

(c) energy = power × time
    energy = 1.8 × 2
    cost = 3.6 × 2 = 72p

The student makes an error at the end but has shown working, so only loses the final mark.

P1b | Radiation and the Universe

Checklist

This spider diagram shows the topics in the unit. You can copy it out and add your notes and questions around it, or cross off each section when you feel confident you know it for your exams.

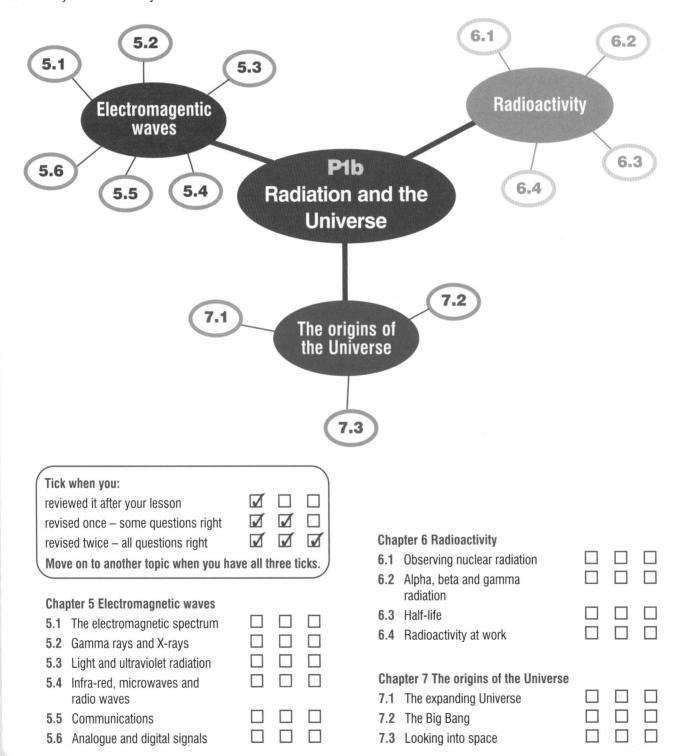

Tick when you:

reviewed it after your lesson	☑	☐	☐
revised once – some questions right	☑	☑	☐
revised twice – all questions right	☑	☑	☑

Move on to another topic when you have all three ticks.

Chapter 5 Electromagnetic waves

5.1	The electromagnetic spectrum	☐	☐	☐
5.2	Gamma rays and X-rays	☐	☐	☐
5.3	Light and ultraviolet radiation	☐	☐	☐
5.4	Infra-red, microwaves and radio waves	☐	☐	☐
5.5	Communications	☐	☐	☐
5.6	Analogue and digital signals	☐	☐	☐

Chapter 6 Radioactivity

6.1	Observing nuclear radiation	☐	☐	☐
6.2	Alpha, beta and gamma radiation	☐	☐	☐
6.3	Half-life	☐	☐	☐
6.4	Radioactivity at work	☐	☐	☐

Chapter 7 The origins of the Universe

7.1	The expanding Universe	☐	☐	☐
7.2	The Big Bang	☐	☐	☐
7.3	Looking into space	☐	☐	☐

What are you expected to know?

Chapter 5 Electromagnetic waves (See students' book pages 78–91)

- Electromagnetic radiation travels as a wave and moves energy from one place to another.

- All electromagnetic waves travel at the same speed through space but have different wavelengths and frequencies.

- The waves together are known as the 'electromagnetic spectrum' and within this they are divided into groups; gamma rays, X-rays, ultraviolet rays, visible light, infra-red rays, microwaves and radio waves.

- When electromagnetic waves move through substances they may be reflected, absorbed or transmitted.

- Different wavelengths of electromagnetic radiation have different effects on living cells.

- There are different uses and hazards for each part of the spectrum.

- Radio waves, microwaves, infra-red and visible light can be used for communication.

- Communication signals can be analogue (continuously varying) or digital (certain values only, usually on or off).

- Electromagnetic waves obey the wave formula:

$$\text{Wave speed} = \text{frequency} \times \text{wavelength}$$

Chapter 6 Radioactivity (See students' book pages 94–103)

- Radioactive substances give out radiation from their nuclei all the time whatever is done to them.

- The three main types of nuclear radiation are alpha particles, beta particles and gamma radiation.

- Properties of alpha particles, beta particles and gamma radiation.

- Half life is the time it takes for the number of parent atoms in a sample to halve.

Chapter 7 The origins of the Universe (See students' book pages 106–113)

- Red shift provides evidence that the Universe is expanding and began with a 'Big Bang'.

- Observations of the Solar System and the galaxies in the Universe can be carried out on the Earth or from space.

- Telescopes used to make observations may detect visible light or other electromagnetic radiations.

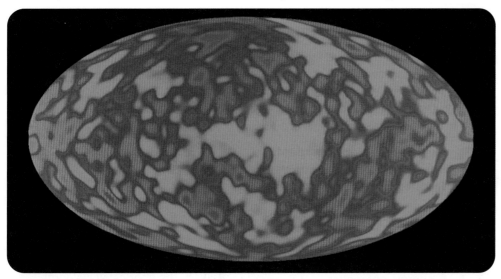

Microwave image of the Universe from COBE, the Cosmic Background Explorer satellite

1. Which part of the electromagnetic spectrum has the longest wavelength?

2. What is the unit of wavelength?

3. How are X-rays used in hospitals?

4. How are gamma rays used in hospitals?

5. How can ultraviolet radiation affect the body?

6. What is visible light?

7. State two uses of microwaves.

8. Which part of the electromagnetic spectrum is detected by night-vision equipment?

9. What is an optical fibre?

10. Which parts of the electromagnetic spectrum are used to carry signals in optical fibres?

11. What is a digital signal?

12. What is meant by 'amplification'?

---

students' book
page 78

P1b 5.1 The electromagnetic spectrum

KEY POINTS

1 The electromagnetic spectrum (in order of increasing wavelength): gamma rays, X-rays, ultraviolet, visible, infra-red, microwaves, radio waves.

2 All electromagnetic waves travel through space at a speed of 300 million m/s

3 Wave = frequency × wave-Speed length
(metres) (hertz, Hz) (metres)

EXAM HINTS

Examination questions often come up about the electromagnetic spectrum. Make sure that you can put the parts of the spectrum in the correct order.

Electromagnetic radiations are electric and magnetic disturbances. They travel as waves and move energy from place to place.

All electromagnetic waves travel through space (a vacuum) at the same speed but they have different wavelengths and frequencies.

All of the waves together are called the 'electromagnetic spectrum'. We group the waves according to their wavelength and frequency:

- Gamma rays have the shortest wavelength and highest frequency.
- Radio waves have the longest wavelength and lowest frequency.
- Different wavelengths of electromagnetic radiation are reflected, absorbed or transmitted differently by different substances and types of surface.

Key words: electromagnetic spectrum, wave speed, wavelength, frequency

GET IT RIGHT!

Electromagnetic waves transfer energy, not matter.

CHECK YOURSELF

1 What is the unit of frequency?

2 Which part of the electromagnetic spectrum has the highest frequency?

3 Which part of the electromagnetic spectrum has the shortest wavelength?

P1b 5.2 Gamma rays and X-rays

KEY POINTS

1 X-rays and gamma radiation are absorbed by dense materials such as bone and metal.
2 X-rays and gamma radiation damage living tissue when they pass through it.
3 X-rays are used in hospitals to take radiographs.
4 Gamma rays are used to kill harmful bacteria in food, to sterilise surgical equipment and to kill cancer cells.

GET IT RIGHT!

Gamma radiation can cause cancer but it is also used to treat cancer.

Gamma rays are used to sterilise surgical instruments and keep food fresh for longer by killing the bacteria on it.

X-rays are used to produce shadow pictures of bones (radiographs).

Gamma rays and X-rays mostly pass through soft body tissues, but some is absorbed and will damage the cells. In lower doses, both radiations can cause changes in the cells that make them cancerous. In higher doses, they can kill the cells.

Gamma radiation is used in hospitals, under carefully controlled conditions, to kill cancer cells.

Working with these radiations is hazardous. Gamma sources are kept in thick lead containers. Staff should wear lead aprons and stand behind lead screens when using X-rays. They monitor their exposure to the radiation with film badges.

Key words: Gamma rays, X-rays, dose, cancer

CHECK YOURSELF

1 How does gamma radiation sterilise surgical instruments?
2 Why do workers in X-ray departments wear lead aprons?
3 Why is a radiograph a 'shadow' picture?

P1b 5.3 Light and ultraviolet radiation

KEY POINTS

1 Ultraviolet radiation is in the electromagnetic spectrum between violet light and X-radiation.
2 Ultraviolet radiation has a shorter wavelength than visible light.
3 Ultraviolet radiation can harm the skin and the eyes.

BUMP UP YOUR GRADE

Make sure you can explain in detail some applications and hazards of ultraviolet radiation.

Ultraviolet radiation has a longer wavelength than X-rays. It has a shorter wavelength than the light at the violet end of the visible spectrum.

Ultraviolet radiation from the Sun causes damage to skin cells – tanning, sunburn, skin ageing and skin cancer. Over-exposure can also damage the eyes. Sun beds work by giving out UV rays.

Fluorescent tubes are coated with substances that absorb the ultraviolet radiation produced inside the tube. Then they emit the energy as visible light.

The same substances are used to make hidden security marks that can only be seen with ultraviolet light.

Visible light is the part of the electromagnetic spectrum that is detected by our eyes. We see the different wavelengths within it as different colours. Visible light can be transmitted along optical fibres.

Key words: ultraviolet, fluorescent, visible, optical fibres

CHECK YOURSELF

1 How can the skin be protected from damage by ultraviolet radiation?
2 How do hidden security markings work?
3 Is the frequency of ultraviolet radiation higher or lower than the frequency of X-rays?

Infra-red, microwaves and radio waves

KEY POINTS

1 **Infra-red:** Heaters, communications (remote handsets, optical fibres)
2 **Microwaves:** Microwave oven, communications
3 **Radio waves:** Communications

 EXAMINER SAYS...

Make sure that you can explain the effects of different wavelengths of electromagnetic radiation on living cells.

BUMP UP YOUR GRADE

Knowing the details here is important. It will gain you extra marks and is the difference between a low-grade and a high-grade answer.

- **Infra-red (IR) radiation** is given out by all objects. The hotter the object, the more IR it emits. Night-vision equipment works by detecting this radiation. IR is absorbed by the skin, we sense it as heat and it can burn. It is used as the heat source in toasters, grills and radiant heaters.
  TV, video and other remote controls use IR. It can be transmitted along optical fibres.
- **Microwaves** are used for cooking and in communications.
  Microwave ovens produce frequencies that are absorbed by water molecules. They heat the water in food, cooking it from the inside out.
  The water in living cells will absorb microwaves and they may be damaged or killed by the heat released.
  Microwave transmitters produce wavelengths that are able to pass through the atmosphere. They are used to send signals to and from satellites and within mobile phone networks.
- **Radio waves** are used to transmit radio and TV programmes.
  When an alternating voltage is applied to an aerial, it emits radio waves with the same frequency as the alternating voltage.
  When the waves are received they produce an alternating current with the same frequency as the radiation.

Key words: infra-red radiation, microwaves, radio waves, communications, optical fibres, transmitters

CHECK YOURSELF

1 How can infra-red cameras be used to find survivors after accidents?
2 Why do microwaves cook food from the inside out?
3 How are radio waves produced?

Communications

KEY POINTS

1 The use we make of radio waves depends on the frequency of the waves.
2 Visible light and infra-red radiation are used to carry signals in optical fibres.

GET IT RIGHT!

Remember that the radio wave part of the spectrum covers a large range of frequencies; from 3000 million Hz to less than 300 000 Hz. The range of wavelengths is 0.1 m to more than 1 km.

The microwave and radio wave part of the electromagnetic spectrum is used for communications. This includes terrestrial TV, satellite TV, mobile phones, emergency services radio, amateur radio transmissions, local, national and international radio.

Different frequencies are used for different applications.

Optical fibres are very thin glass fibres. They are flexible and can be bent around curves. Light or infra-red radiation is transmitted along the fibre by continuous reflections.

Key words: communications, optical fibres, reflections

CHECK YOURSELF

1 Which electromagnetic waves are used for transmissions along optical fibres?
2 How are electromagnetic waves transmitted along optical fibres?
3 Which type of electromagnetic wave is used for satellite communications?

KEY POINTS

1 Analogue signals vary continuously in amplitude.
2 Digital signals are either high ('1') or low ('0').
3 Digital transmission, when compared with analogue transmission, is free of noise and distortion. It can also carry much more information.

GET IT RIGHT!

In this context, 'noise' means unwanted, usually random, impulses that get added to the original signal.

Communication signals are either analogue or digital.

- An analogue signal varies continuously in amplitude.
- Digital signals only have certain values. Usually they are either high (on/'1') or low (off/'0'). They can be processed by computers.

When signals are transmitted they become less strong over distance and have to be amplified. They also pick up noise. When amplification takes place the noise is also amplified. With analogue this can make the signal very distorted. With digital the signal can be 'cleaned', because it is still clear which part of the signal is high and which part is low.

Key words: analogue, digital, noise, distortion, amplified

EXAMINER SAYS...

Remember that digital signals can carry much more information than analogue signals. Digital pulses can be made very short so that many pulses can be transmitted each second.

CHECK YOURSELF

1 Why do signals need to be amplified?
2 Why do analogue signals get distorted when they are amplified?
3 Which type of signal can be processed by a computer?

P1b 5   End of chapter questions

1 What are electromagnetic waves?

2 What do all electromagnetic waves have in common?

3 Which part of the electromagnetic spectrum has the lowest frequency?

4 What are gamma rays used for?

5 What is infra-red radiation used for in the home?

6 What is an analogue signal?

7 How are workers in radiography departments in hospitals protected from exposure to X-rays?

8 How does the wavelength of ultraviolet radiation compare with the wavelength of visible light?

9 Why are microwaves dangerous to the body?

10 How does the frequency of radio waves compare with the frequency of visible light?

11 How are optical fibres used in medicine?

12 What is the formula that relates frequency, speed and wavelength?

1. Describe the basic structure of a nucleus.

2. What is the effect of pressure on the rate of radioactive decay?

3. What is the structure of an alpha particle?

4. What is the range of alpha particles in air?

5. What happens to the radioactivity of a sample of a radioactive material over time?

6. What happens to the count rate from a radioactive sample during one half life?

7. Where are alpha sources commonly found in the home?

8. Which is the most dangerous type of nuclear radiation if the source is inside the body?

---

students' book page 94

P1b 6.1   Observing nuclear radiation

KEY POINTS

1. A radioactive substance contains unstable nuclei.
2. An unstable nucleus becomes stable by emitting radiation.
3. There are three types of radiation from radioactive substances – alpha, beta and gamma radiation.
4. Radioactive decay is a random event – we cannot predict or influence when it will happen.

The basic structure of an atom is a small central nucleus, made up of protons and neutrons, surrounded by electrons.

The atoms of an element always have the same number of protons. However, different isotopes of the element will have different numbers of neutrons.

The nuclei of radioactive substances are unstable. They become stable by radioactive decay. In this process, they emit radiation and turn into other elements.

The three types of radiation emitted are:

- alpha particles
- beta particles
- gamma rays.

Radioactive decay is a random process and is not affected by external conditions.

Key words: nuclei, proton, neutron, radioactive decay, alpha particles, beta particles, gamma rays

GET IT RIGHT!

Radioactive decay is a random process. It is not possible to predict when any particular nucleus will decay and it is not possible to make any particular nucleus decay.

Radioactive decay is not affected by external conditions. So you cannot make it happen faster by changing things like temperature or pressure.

CHECK YOURSELF

1. Which part of an atom might emit alpha particles?

2. What is meant by 'isotopes of an element'?

3. What happens to the rate of radioactive decay if the temperature is doubled?

P1b 6.2 Alpha, beta and gamma radiation

KEY POINTS

1 α-radiation is stopped by paper or a few centimetres of air.
2 β -radiation is stopped by thin metal or about a metre of air.
3 γ -radiation is stopped by thick lead and has an unlimited range in air.

EXAM HINTS

For each of the three types of radiation you need to remember:
● what it is
● how ionising it is
● how far it penetrates different materials
● if it is deflected by electric and magnetic fields.

● **An alpha (α) particle** is a helium nucleus. It is made up of 2 protons and 2 neutrons.
● **A beta (β) particle** is a high-speed electron from the nucleus. It is emitted when a neutron changes to a proton and an electron. The proton remains in the nucleus.
● **Gamma (γ) radiation** is very short wavelength electromagnetic radiation that is emitted from the nucleus.

When nuclear radiation travels through a material it will collide with the atoms of the material and knock electrons off them, creating ions. This is called 'ionisation'.

Alpha particles are relatively large, so they have lots of collisions with atoms – they are strongly ionising. Because of these collisions, the alpha particles do not penetrate far into a material. They can be stopped by a thin sheet of paper, human skin or a few centimetres of air. Alpha particles have a positive charge and are deflected by electric and magnetic fields.

Beta particles are much smaller and faster than alpha particles so they are less ionising and penetrate further. They are blocked by a few metres of air or a thin sheet of aluminium. Beta particles have a negative charge and are deflected by electric and magnetic fields in the opposite sense to alpha particles.

Gamma rays are electromagnetic waves so they will travel a long way through a material before colliding with an atom. They are weakly ionising and very penetrating. Several centimetres of lead or several metres of concrete are needed to absorb most of the radiation. Gamma rays are not deflected by electric and magnetic fields.

Key words: ionisation, electric and magnetic fields, charge

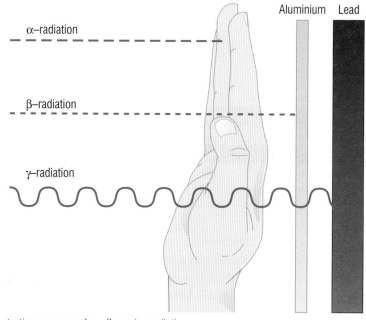

The penetrating powers of α-, β- and γ-radiation

CHECK YOURSELF

1 Which type of nuclear radiation is the most penetrating?

2 Which type of nuclear radiation is the most ionising?

3 Why is gamma radiation not deflected by electric and magnetic fields?

P1b 6.3 Half-life

The half-life of a radioactive substance is the time it takes:

1 for the number and (therefore the mass) of parent atoms in a sample to halve
2 for the count rate from the original substance to fall to half its initial level.

BUMP UP YOUR GRADE

Practise doing half-life calculations.

We can measure the radioactivity of a sample of a radioactive material by measuring the count rate from it.

The radioactivity of a sample decreases over time. How quickly the count rate falls to nearly zero depends on the material. Some take a few minutes, others take millions of years.

We use the idea of half-life to measure how quickly the radioactivity decreases. It is the time taken for the count rate from the original substance to fall to half its initial value.

Or we can define it as the time it takes for the number of unstable nuclei in a sample to halve.

The half-life is the same for any sample of a particular material.

Key words: half-life, count rate

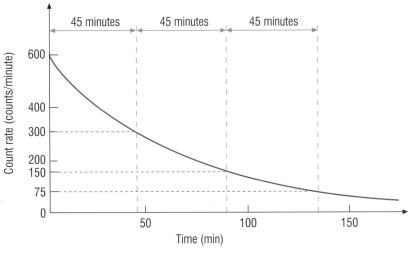

Radioactive decay: a graph of count rate against time

1 What happens to the count rate from a sample over time?

2 What has happened to the original count rate of a sample after two half-lives have passed?

3 What has happened to the number of atoms in a sample after two half-lives have passed?

P1b 6.4 Radioactivity at work

The use we can make of a radioactive substance depends on:

1 its half-life, and
2 the type of radiation it gives out.

- **Alpha sources** are used in smoke alarms. The alpha particles are not dangerous because they are very poorly penetrating. The source needs a half life of several years.
- **Beta sources** are used for thickness control in the manufacture of things like paper. Alpha particles would be stopped by a thin sheet of paper and all gamma rays would pass through it. The source needs a half life of many years, so that decreases in count rate are due to changes in the thickness of the paper. (See diagram on the next page.)

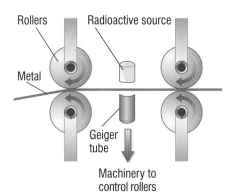

Rollers Radioactive source

Metal

Geiger tube

Machinery to control rollers

Thickness monitoring using a radioactive source

• **Gamma and beta sources** are used as 'tracers' in medicine. The source is injected or swallowed by the patient. Its progress around the body is monitored by a detector outside the patient. The source needs a half life of a few hours so that the patient is not exposed to unnecessary radioactivity.

• **Gamma sources** are also used to sterilise medical equipment and prevent food spoilage.

If nuclear radiation enters living cells, it causes ionisation which damages cells and may cause cancer.

If the source of radiation is outside the body, alpha particles will be stopped by clothing or skin. Gamma and beta radiation are more dangerous because they may pass through skin and damage cells.

If the source of radiation is inside the body (e.g. it is inhaled), alpha radiation is the most dangerous because it is very strongly ionising.

Key words: tracers

BUMP UP YOUR GRADE

For each radiation you should know an application, why a particular source is used and the approximate half life.

CHECK YOURSELF

1 Why is a beta source less dangerous inside the body than an alpha source?

2 Why do medical tracers have half lives of just a few hours?

3 Why isn't an alpha source used as a tracer in medicine?

P1b 6 — End of chapter questions

1 **What is a beta particle?**

2 **What is the range of beta particles in air?**

3 **What is meant by 'half-life'?**

4 **What is gamma radiation?**

5 **Which is the most dangerous type of nuclear radiation if the source is outside the body?**

6 **What happens to the count rate from a radioactive sample during three half-lives?**

7 **Which type of nuclear radiation is the least ionising?**

8 **Which type of nuclear radiation is used for thickness control in the manufacture of paper?**

1. What happens to the sound waves we hear if the source of the waves is moving?

2. Which galaxy is our Sun part of?

3. What happens to light from distant galaxies before it reaches the Earth?

4. How do most scientists think the Universe started?

5. What is a telescope?

6. Where are telescopes used?

students' book
page 106

P1b 7.1 — The expanding Universe

KEY POINTS

1. Light from a distant galaxy is red-shifted to longer wavelengths.
2. The further away the galaxy the bigger the red shift.

Galaxies

We live in a galaxy called the Milky Way.

It contains billions of stars

But there are billions of galaxies in the Universe so it is hard to imagine the total number of stars.

GET IT RIGHT!

The further away from us a galaxy is, the faster it is moving away from us.

Red shift

If a source of waves is moving relative to an observer, the wavelength and frequency 'seen' by the observer will have changed (shifted) from the original produced by the source.

This effect can be heard with sound waves. For example the sound of an ambulance siren will sound different depending on whether it is moving towards you (pitch is higher) or away from you (pitch is lower).

The effect occurs with light waves. Light observed from distant galaxies has been 'shifted' towards the red end of the spectrum. This means the frequency has decreased. The further away the galaxy, the bigger the red shift.

This suggests that distant galaxies are moving away from us, and the most distant galaxies are moving the fastest. This is true of galaxies no matter which direction you look in.

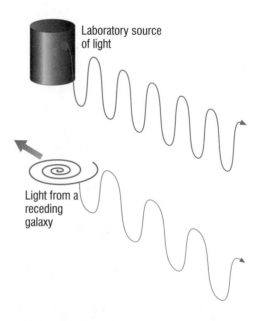

Laboratory source of light

Light from a receding galaxy

Red shift

- Galaxies are collections of billions of stars.
- Our Sun is one of the stars in the Milky Way galaxy.
- The Universe is made up of billions of galaxies.

Key words: red shift, galaxies

CHECK YOURSELF

1 What is a 'galaxy'?

2 What is 'red shift'?

3 Which galaxies are moving away from us fastest?

P1b 7.2 The Big Bang

KEY POINTS

1 Red shift provides evidence that the Universe is expanding.
2 The Universe started with the Big Bang, a massive explosion from a very small point.

EXAMINER SAYS...

Be sure you can explain why red shift is evidence for an expanding Universe and the Big Bang.

Red shift shows us that distant galaxies are moving away from us and the furthest ones are moving the fastest. This gives us evidence that the Universe is expanding outwards in all directions.

We can try and imagine backwards in time to see where the Universe came from. If it is now expanding outwards, this suggests that it started with a massive explosion at a very small initial point. This is known as the 'Big Bang' theory.

Key words: expanding, Big Bang

The Big Bang

CHECK YOURSELF

1 How does red shift show us that the Universe is expanding?

2 What is meant by the 'Big Bang'?

3 How is an expanding Universe evidence for the Big Bang?

P1b 7.3 Looking into space

KEY POINTS

1 Observations are made with telescopes that may detect visible light or other electromagnetic radiations.
2 Observations of the Solar System and galaxies can be carried out from the Earth or from space.

GET IT RIGHT!

Not all telescopes detect visible light. Some 'see' into space by detecting radiation from other parts of the electromagnetic spectrum.

Scientists use telescopes to collect the visible light coming from stars, and so see them. They can also use telescopes that collect radiation from other parts of the electromagnetic spectrum such as X-rays, or radio waves. This also allows them to 'see' distant stars.

The atmosphere is a layer of gases surrounding the Earth.

Telescopes on satellites are able to receive all types of electromagnetic radiation from space, without it being distorted or absorbed by the Earth's atmosphere. Because there is no distortion, the pictures produced by these telescopes are clearer and have more detail. So it is possible for us to observe stars that are further away.

A comet

Key words: telescope, atmosphere, satellite

CHECK YOURSELF

1 What is the Earth's atmosphere?

2 What does the Earth's atmosphere do to electromagnetic radiation from space?

3 Why are telescopes on satellites able to produce clearer pictures than those on Earth?

P1b 7 End of chapter questions

1 **Roughly how many galaxies is the Universe thought to contain?**

2 **What would a 'blue shift' of light from distant galaxies show?**

3 **What evidence is there that the Universe is expanding?**

4 **What has happened to the frequency of light that reaches the Earth from distant galaxies?**

5 **What is the advantage of using a telescope on a satellite rather than on Earth?**

6 **What is a radio telescope?**

1 Different parts of the electromagnetic spectrum have different uses.

(a) Which electromagnetic radiation is used:
(i) to treat cancer?
(ii) to cook food from the inside? (2 marks)

(b) Ultraviolet radiation can be both useful and hazardous:
(i) What effects can ultraviolet radiation have on the skin? (4 marks)
(ii) Valuable items can be marked with a special ink. The ink can only be seen when ultraviolet radiation is shone on it.

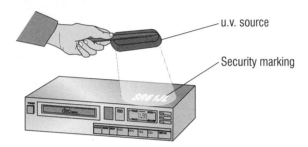

Explain how the ink becomes visible in ultraviolet radiation. (2 marks)

(c) State two things that all parts of the electromagnetic spectrum have in common. (2 marks)

2 The activity of a radioactive source is measured. At one instant it is 120 counts per second. 180 seconds later it is 15 counts per second.

(a) How many half-lives are there between these two measurements? (2 marks)

(b) Calculate the half-life of the radioactive source. (2 marks)

(c) A more reliable value for the half-life could be found by plotting a graph of activity against time. Explain what measurements you would make in order to plot the graph. (2 marks)

(d) Why would the half-life calculated from the graph be more reliable? (1 mark)

3 Different parts of the electromagnetic spectrum are used in hospitals for different applications.

Describe as fully as you can the following applications.

(a) Using gamma rays to ensure medical equipment is clean. (3 marks)

(b) Using gamma rays for the treatment of cancer. (3 marks)

(c) Using X rays to produce radiographs. (3 marks)

4 There are three types of nuclear radiation, alpha, beta and gamma.

(a) Which types of nuclear radiation, can be deflected by electric and magnetic fields? Explain your answer. (3 marks)

(b) When nuclear radiations pass through materials they cause different amounts of ionisation.
(i) Explain what is meant by ionisation. (2 marks)
(ii) Which nuclear radiation is the most ionising? Explain your answer. (3 marks)

5 (a) Radio telescopes and optical telescopes can both be used to investigate distant galaxies.
(i) Why does an optical telescope allow you to see stars that are too faint to see just with your eyes? (1 mark)
(ii) What is a radio telescope? (1 mark)
(iii) Why is a radio telescope larger than an optical telescope? (1 mark)
(iv) Explain why a radio telescope can be used on the ground even in places where it would be impossible to use an optical telescope. (2 marks)

(b) Some objects in space produce gamma radiation. Why does a gamma ray detector need to be on a satellite? (2 marks)

Test & Assessment Interactive quizzes, answers and hints online!

The answer is worth 4 marks out of the 6 available. The responses worth a mark are underlined in red.

We can improve the answers in several ways:

(a) (i) Why is an alpha particle emitter with a half life of 60 seconds *not* suitable for use in a smoke alarm? *(1 mark)*

(ii) Why are beta and gamma sources *not* suitable for use in a smoke alarm? *(2 marks)*

(b) Radioactive sources that emit beta particles are used in the manufacture of paper to monitor the thickness of the paper. Explain why a beta emitter is suitable for this, but alpha or gamma emitters are not. *(3 marks)*

(a)(i) Alpha particle sources are used in smoke alarms, but the half life of this one is so short that the alarm would hardly last any time.

(ii) They are too penetrating.

(b) Gamma rays would just go straight through the paper without being stopped.
Alpha particles would not go through it at all.

An extra mark would be given for stating that **they might affect people near the smoke alarm**.

The extra mark would be given for explaining why a beta emitter is suitable. This is because **the number of beta particles which pass through the paper varies according to the thickness of the paper**.

The answer is worth 2 marks out of the 5 available. The responses worth a mark are underlined in red.

We can improve the answer in several ways:

X-rays are both useful and hazardous:

(a) Give a use for X-rays. *(1 mark)*

(b) Why are X-rays hazardous? *(1 mark)*

(c) What measures can be taken to prevent people who work with X-rays from being exposed to too much radiation? *(3 marks)*

(a) X-rays

(b) they are dangerous

(c) wear lead aprons and stand behind lead screens

The student may mean that a use for X-rays is to take X-ray pictures, but this is not clear from the answer, so there is no mark. She should have explained that **X-rays are used to take shadow pictures of bones (radiographs)**.

This answer is also too vague to score a mark. The student should have said that **X-rays can damage or kill cells**.

Two means of protection are given. An extra mark could be gained by stating that **film badges are worn to monitor exposure to X-rays** (although they don't in themselves protect the wearer).

P2 | Additional physics (Chapters 1–3)

Checklist

This spider diagram shows the topics in the first half of the unit. You can copy it out and add your notes and questions around it, or cross off each section when you feel confident you know it for your exams.

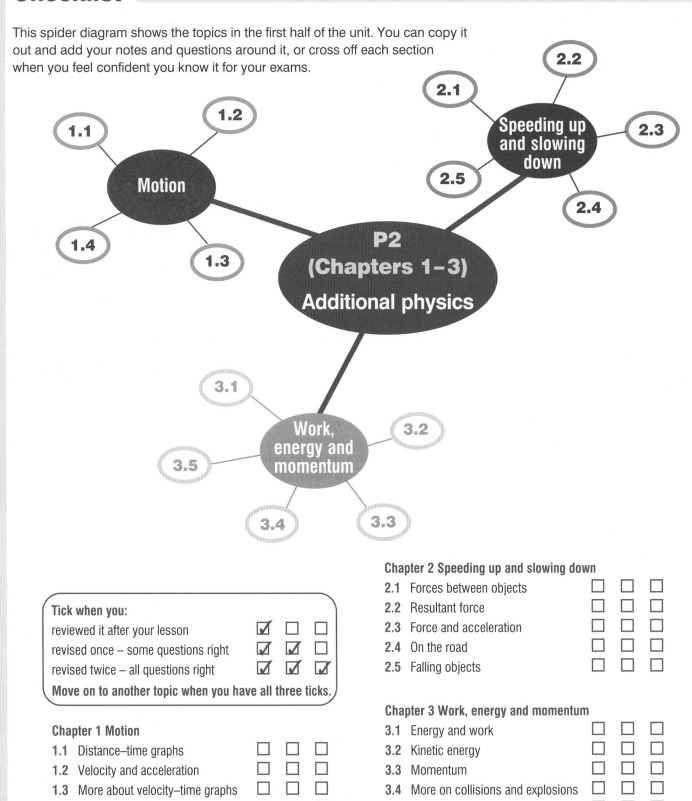

Tick when you:

reviewed it after your lesson	☑	☐	☐
revised once – some questions right	☑	☑	☐
revised twice – all questions right	☑	☑	☑

Move on to another topic when you have all three ticks.

Chapter 1 Motion

1.1	Distance–time graphs	☐	☐	☐
1.2	Velocity and acceleration	☐	☐	☐
1.3	More about velocity–time graphs	☐	☐	☐
1.4	Using graphs	☐	☐	☐

Chapter 2 Speeding up and slowing down

2.1	Forces between objects	☐	☐	☐
2.2	Resultant force	☐	☐	☐
2.3	Force and acceleration	☐	☐	☐
2.4	On the road	☐	☐	☐
2.5	Falling objects	☐	☐	☐

Chapter 3 Work, energy and momentum

3.1	Energy and work	☐	☐	☐
3.2	Kinetic energy	☐	☐	☐
3.3	Momentum	☐	☐	☐
3.4	More on collisions and explosions	☐	☐	☐
3.5	Changing momentum	☐	☐	☐

What are you expected to know?

Chapter 1 Motion _(See students' book pages 120–129)_

- Construction and use of distance–time graphs.
- Construction and use of velocity–time graphs.
- Acceleration = change in speed/time taken for the change

Chapter 2 Speeding up and slowing down _(See students' book pages 132–143)_

- Weight = mass × gravitational field strength
- What is meant by a resultant force?
- Resultant forces produce accelerations.
- Resultant force = mass × acceleration
- Factors that affect stopping distance.
- Factors that affect reaction time and braking distance.
- Frictional forces oppose motion.
- Bodies falling through fluids reach a terminal velocity.

Chapter 3 Work, energy and momentum _(See students' book pages 146–157)_

- Work done = energy transferred
- Work done = force × distance moved in the direction of the force
- Kinetic energy = $\frac{1}{2}$ mass × speed^2
- Momentum = mass × velocity
- Total momentum is conserved in a collision or explosion provided no external forces act.
- Force = change in momentum/time taken for the change

(1) **What is the SI unit of speed?**

(2) **What does the distance–time graph for a stationary body look like?**

(3) **What is the equation that allows us to calculate acceleration?**

(4) **What does a negative value for acceleration mean?**

(5) **What does the slope of the line on a velocity–time graph represent?**

(6) **What would the velocity–time graph for a body moving at a constant speed look like?**

(7) **How do you calculate the slope of a line on a graph? [Higher Tier only]**

(8) **What happens to the slope of the line on a distance–time graph if the speed increases?**

---

students' book
page 120

P2 1.1 Distance–time graphs

KEY POINTS

1 The steeper the line on a distance–time graph, the greater the speed it represents.

2 Speed (metre/second, m/s) =

$$\frac{\text{distance travelled (metre, m)}}{\text{time taken (second, s)}}$$

BUMP UP YOUR GRADE

Always show the stages in your working when you do calculations. Always include a unit with your answer if one is not given.

We can use graphs to help us describe the motion of a body.

A distance–time graph shows the distance of a body from a starting point (y-axis) against time taken (x-axis).

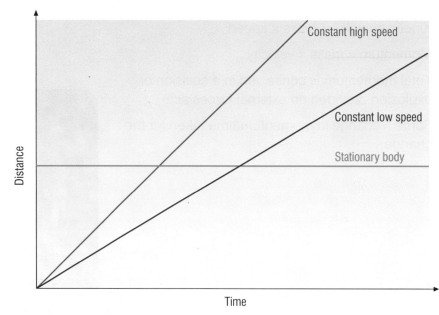

Distance–time graphs

GET IT RIGHT!

Make sure that you always label the axes on a graph with a quantity and a unit.

The slope of the line on a distance–time graph represents speed.

The steeper the slope, the greater the speed.

The speed of a body is the distance travelled each second. We can calculate the speed of a body using the equation:

$$speed = \frac{distance\ travelled}{time\ taken}$$

The SI unit of speed is metres per second (m/s).

Key words: graph, distance, time, metre, second, slope

CHECK YOURSELF

1 What does the slope of the line on a distance–time graph represent?

2 What is the equation that relates speed, distance and time?

3 What is the SI unit of distance?

---

students' book
page 122

P2 1.2 Velocity and acceleration

KEY POINTS

1 Velocity is speed in a given direction.
2 Acceleration is change of velocity per second.
3 A body travelling at a steady speed is accelerating if its direction is changing.

GET IT RIGHT!

Remember that a velocity must include a direction.

If a body changes direction, it changes velocity. So it accelerates, even if its speed stays the same.

EXAM HINTS

There are several units to learn here. Take care not to confuse m/s (unit of speed and velocity) and m/s^2 (unit of acceleration).

The velocity of a body is its speed in a given direction. If the body changes direction it changes velocity, even if its speed stays the same.

If the velocity of a body changes, we say that it accelerates.

We can calculate acceleration using the equation:

$$acceleration = \frac{change\ in\ velocity}{time\ taken\ for\ the\ change}$$

The SI unit of acceleration is metres per second squared (m/s^2).

If the value calculated for acceleration is negative, the body is decelerating – slowing down.

Key words: velocity, direction, acceleration, decelerating

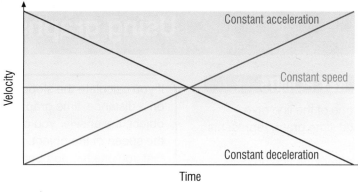

Velocity–time graphs

CHECK YOURSELF

1 What is the difference between speed and velocity?

2 What is the SI unit of acceleration?

3 What do we mean by deceleration?

P2 1.3 More about velocity–time graphs

KEY POINTS

1 The slope of a line on a velocity–time graph represents acceleration.
2 The area under the line on a velocity–time graph represents distance travelled.

AQA EXAMINER SAYS...

Take care not to confuse distance–time graphs and velocity–time graphs.

A velocity–time graph shows the velocity of a body (*y*-axis) against time taken (*x*-axis).

- The slope of a line on a velocity–time graph represents acceleration.
- The steeper the slope, the greater the acceleration.
- If the slope is negative, the body is decelerating.
- The area under the line on a velocity–time graph represents the distance travelled in a given time.
- The bigger the area, the greater the distance travelled.

Key words: area

CHECK YOURSELF

1 What does a horizontal line on a velocity–time graph represent?
2 What would the velocity–time graph for a steadily decelerating body look like?
3 What does the area under the line on a velocity–time graph represent?

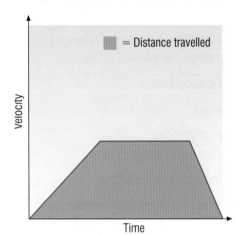

Area under a velocity–time graph

P2 1.4 Using graphs

KEY POINTS

1 The slope of the line on a distance–time graph represents speed.
2 The slope of the line on a velocity–time graph represents acceleration.
3 The area under the line on a velocity–time graph represents the distance travelled.
   [Calculations of slopes are required for Higher Tier only]

If you calculate the slope of the line on a distance–time graph for an object, the answer you obtain will be the speed of the object. Make sure that you use the numbers from the graph scales in your calculations.

If you calculate the slope of the line on a velocity–time graph for an object, the answer you obtain will be the acceleration of the body. Make sure that you use the numbers from the graph scales in your calculations.

HIGHER

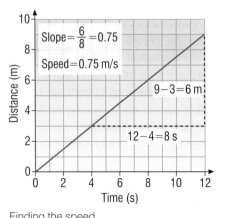

Finding the speed

Calculating the area under the line on a velocity–time graph between two times gives the distance travelled between those times.

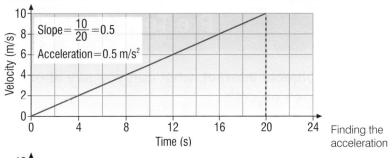

Slope $= \frac{10}{20} = 0.5$

Acceleration $= 0.5$ m/s^2

Finding the acceleration

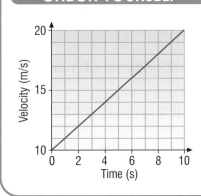

Area under graph

$= \frac{1}{2}(20 \times 8) + (40 \times 8)$

$= \frac{1}{2} \times 160 + 320$

$= 80 + 320$

$= 400$ m

Finding distance travelled

Key words: graph scales

GET IT RIGHT!

If the line on a velocity–time graph for a body has a negative slope, the body is decelerating.

EXAM HINTS

Practise calculating slopes of lines on graphs and the area under the lines. Make sure that you use values that you have read from the graph scales.

CHECK YOURSELF

The graph shows the motion of a car.

1 What is the initial speed of the car?

2 What is the final speed of the car?

3 What is the acceleration of the car?

P2 1   End of chapter questions

1 **What does a horizontal line on a distance–time graph represent?**

2 **What is the average speed of a sprinter who runs 100 m in 16 s?**

3 **How can a body travelling at a steady speed be accelerating?**

4 **What quantity has the unit m/s²?**

5 **What does a negative slope on a velocity–time graph mean?**

6 **What part of a velocity–time graph represents distance travelled?**

7 **Look at the top graph. What is the speed of the runner?**

8 **Look at the bottom graph. What is the distance travelled by the car in the first 10 seconds? [Higher Tier only]**

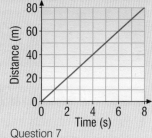

Question 7

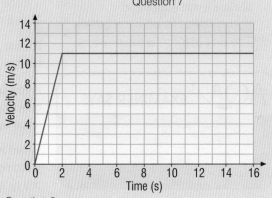

Question 8

1. When two objects interact, what can you say about the size of the forces acting on each object?

2. If you stand on flat ground, what is the direction of the reaction force from the ground on you?

3. What is a resultant force?

4. What effect does a resultant force have on a stationary object?

5. What is the equation that relates mass, resultant force and acceleration?

6. What acceleration is produced when a resultant force of 15 N acts on a mass of 5 kg?

7. What do we mean by 'the reaction time' of a driver?

8. What factors might increase the braking distance of a car?

9. What do we mean by 'terminal velocity'?

10. What is the equation that relates gravitational field strength, mass and weight?

---

students' book page 132

P2 2.1   Forces between objects

KEY POINTS

1. When two objects interact, they always exert equal and opposite forces on each other.
2. The unit of force is the newton.

AQA EXAMINER SAYS...

Remember that action and reaction forces act on different objects.

Forces are measured in newtons, abbreviated to N.

Objects always exert equal and opposite forces on each other. If object A exerts a force on object B, object B exerts an equal and opposite force on object A. These are sometimes called 'action and reaction' forces.

If a car hits a barrier it exerts a force on the barrier. The barrier exerts a force on the car that is equal in size and in the opposite direction.

If you place a book on a table the weight of the book will act vertically downwards on the table. The table will exert an equal and opposite reaction force upwards on the book.

When a car is being driven forwards there is a force from the tyre on the ground pushing backwards. There is an equal and opposite force from the ground on the tyre which pushes the car forwards.

Key words: force, newton, equal, opposite

Pull ←   Pull →

Equal and opposite forces

GET IT RIGHT!

Forces have both size and direction.

CHECK YOURSELF

1 What abbreviation is used for the unit of force?

2 In which direction does the force of weight always act?

3 If you push on a wall with a horizontal force of 15 N to the right, what force will the wall exert on you?

students' book page 134

P2 2.2   Resultant force

KEY POINTS

	Object at the start	Resultant force	Effect on the object
1	at rest	zero	stays at rest
2	moving	zero	velocity stays the same
3	moving	non-zero in the same direction as the direction of motion of the object	accelerates
4	moving	non-zero in the opposite direction to the direction of motion of the object	decelerates

Most objects have more than one force acting on them. The 'resultant force' is the single force that would have the same effect on the object as all the original forces acting together.

When the resultant force on an object is zero:

- if the object is at rest it will stay at rest
- if the object is moving it will carry on moving at the same speed and in the same direction.

GET IT RIGHT!

If an object is accelerating, there must be a resultant force acting on it.

When the resultant force on an object is not zero there will be an acceleration in the direction of the force.

This means that:

- if the object is at rest it will accelerate in the direction of the resultant force
- if the object is moving in the same direction as the resultant force it will accelerate in that direction
- if the object is moving in the opposite direction to the resultant force it will decelerate.

Key words: resultant force, accelerate, decelerate

AQA EXAMINER SAYS...

If an object is accelerating, it can be speeding up, slowing down or changing direction.

CHECK YOURSELF

1 What happens to an object moving at a steady speed if the resultant force on it is zero?

2 What are the units of resultant force?

3 When will a resultant force cause a deceleration?

P2 2.3 Force and acceleration

KEY POINTS

Resultant force (newtons, N) = mass (kilograms, kg) × acceleration (metres/second², m/s²)

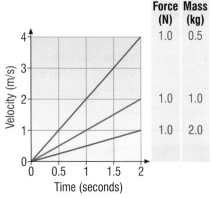

	Force (N)	Mass (kg)
	1.0	0.5
	1.0	1.0
	1.0	2.0

Different combinations of force and mass

A resultant force always causes an acceleration. Remember that a deceleration is a negative acceleration.

If there is no acceleration in a particular situation the resultant force must be zero.

Acceleration is a change in velocity. An object can accelerate by changing its direction even if it is going at a constant speed. So a resultant force is needed to make an object change direction.

Resultant force, mass and acceleration are related by the equation:

$$\text{resultant force} = \text{mass} \times \text{acceleration}$$

The greater the resultant force on an object, the greater its acceleration.

The bigger the mass of an object, the bigger the force needed to give it a particular acceleration.

Key words: **resultant force, mass, acceleration, deceleration**

CHECK YOURSELF

1 What is the equation that links mass, acceleration and resultant force?

2 What happens to the acceleration of an object as the resultant force on it increases?

3 What is the acceleration of a car of mass 2000 kg if the resultant force acting in its direction of motion is 8 N?

P2 2.4 On the road

KEY POINTS

1 The 'thinking' distance is the distance travelled by the vehicle in the time it takes the driver to react.
2 The braking distance is the distance the vehicle travels under the braking force.
3 Stopping distance = thinking distance + braking distance

GET IT RIGHT!

The reaction time depends on the driver. The braking distance depends on the road and weather conditions and the condition of the vehicle.

If a vehicle is travelling at a steady speed the resultant force on it is zero.

The driving forces are equal and opposite to the frictional forces.

The faster the speed of a vehicle, the bigger the deceleration needed to bring it to rest in a particular distance. So the bigger the braking force needed.

The stopping distance of a vehicle is the distance it travels during the driver's reaction time (the thinking distance) plus the distance it travels under the braking force (the braking distance).

The thinking distance is increased if the driver is tired or under the influence of alcohol or drugs.

The braking distance can be increased by poorly maintained roads, bad weather conditions and the condition of the car. For example, worn tyres or worn brakes will increase braking distance.

Key words: **stopping distance, thinking distance, braking distance**

CHECK YOURSELF

1 What is the resultant force on a car travelling at a steady speed on a straight horizontal road?

2 What is the relationship between stopping distance, thinking distance and braking distance?

3 What is the effect of the speed of a vehicle on its stopping distance?

P2 2.5 Falling objects

KEY POINTS

1 The weight of an object is the force of gravity on it.
2 An object falling freely accelerates at about $10\,m/s^2$.
3 An object falling in a fluid reaches a terminal velocity.

GET IT RIGHT!

Do not confuse weight and mass. Remember that weight is the force of gravity acting on an object. Mass is the amount of matter in an object.

AQA EXAMINER SAYS...

The resistive force exerted by a fluid is sometimes called the 'drag force'.

If an object falls freely, the resultant force acting on it is the force of gravity. It will make the object accelerate at about $10\,m/s^2$ close to the Earth's surface.

We call the force of gravity 'weight', and the acceleration 'the acceleration due to gravity'.

The equation   resultant force = mass × acceleration

becomes   weight (N) = mass (kg) × acceleration due to gravity (m/s²)

If the object is on the Earth, not falling, we calculate the weight using:

weight (N) = mass (kg) × gravitational field strength (N/kg)

When an object falls through a fluid (e.g. air), the fluid exerts frictional forces (e.g. air resistance) on it, resisting its motion. The faster the object falls, the bigger the frictional force becomes. Eventually it will be equal to the weight of the object. The resultant force is now zero, so the body stops accelerating and moves at a constant velocity called the 'terminal velocity'.

We can show the motion of the object on a velocity–time graph.

Key words: weight, acceleration due to gravity, gravitational field strength, terminal velocity

CHECK YOURSELF

1 What are the units of gravitational field strength?

2 Why does an object dropped in a fluid initially accelerate?

3 What eventually happens to an object falling through a fluid?

P2 2 End of chapter questions

1 **If you pull on a rope attached to a wall with a force of 5 N, with what force will the rope pull on you?**

2 **When two objects interact, what can you say about the directions of the forces acting?**

3 **What effect does a resultant force have on the motion of an object moving in the same direction as the force?**

4 **What effect does a resultant force have on the motion of an object moving in the opposite direction to the force?**

5 **What happens to a body moving in a straight line if the resultant force acting on it increases?**

6 **What is the resultant force on a body of mass 70 kg accelerating at $2\,m/s^2$?**

7 **What do we mean by the 'thinking distance' of a car driver?**

8 **What factors might increase thinking distance?**

9 **What is the force of gravity acting on a object called?**

10 **What is the resultant force acting on a object at terminal velocity?**

1. What is work done equivalent to?

2. What is the equation relating force, work done and distance?

3. What is the energy of movement called?

4. What is elastic potential energy?

5. What is the equation relating mass, velocity and momentum?

6. In what situations does the conservation of momentum apply?

7. A gun of mass 1 kg fires a bullet of mass 0.005 kg at 100 m/s. What is the recoil velocity of the gun?

8. A white snooker ball of mass 0.1 kg moving at 0.5 m/s hits an identical, stationary red ball. After the collision, the white ball continues in the same direction at a speed of 0.2 m/s. What is the speed of the red ball after the collision?

9. How does a crumple zone reduce the forces on a car in an accident?

10. A force of 2000 N is needed to bring a car to rest in 8 s. What is the momentum of the car?

---

students' book page 146

P2 3.1   Energy and work

KEY POINTS

1 Work done = energy transferred
2 Work done (joules, J) = force (newtons, N) × distance moved in the direction of the force (metres, m)

GET IT RIGHT!

Remember that work done is equal to energy transferred.

When a force moves an object, energy is transferred and work is done.

Whenever an object starts to move, a force must have been applied to it. This force needs a supply of energy from somewhere, such as electricity or fuel. When work is done moving the object, the supplied energy is transferred to the object so the work done is equal to the energy transferred.

Both work and energy have the unit joule, J.

When work is done against frictional forces, the energy supplied is mainly transformed into heat.

The work done on an object is calculated using the equation:

$$\text{work done} = \text{force} \times \text{distance moved in the direction of the force}$$

Notice that if the distance moved is zero, no work is done on the object.

Key words: energy transferred, work done, joule

CHECK YOURSELF

1 What is the unit of work done?

2 When is work done by a force?

3 What work is done on an object if a force of 300 N moves it a distance of 7 m?

P2 3.2 Kinetic energy

KEY POINTS

1 Elastic potential energy is the energy stored in an elastic object when work is done on the object.
2 The kinetic energy of an object depends on its mass and its speed.
3 Kinetic energy (joule, J)
$= \frac{1}{2} \times$ mass (kilogram, kg)
$\times$ speed^2 (metre/second)^2
[Higher Tier only]

BUMP UP YOUR GRADE

Take care when you use the kinetic energy equation. In an examination question students often forget to square the speed.

An elastic object is one that will go back to its original shape after it has been stretched or squashed.

When work is done on an elastic object to stretch or squash it, the energy transferred to it is stored as elastic potential energy. When the object returns to its original shape this energy is released.

Kinetic energy is the energy of movement.

The kinetic energy of a body depends on its mass and its speed. The greater its mass and the faster its speed, the more kinetic energy it has.

Kinetic energy can be calculated using the equation:

$$\text{kinetic energy} = \tfrac{1}{2}\text{mass} \times \text{speed}^2$$

CHECK YOURSELF

1 What happens to the kinetic energy of an object as its mass increases?
2 What are the units of kinetic energy?
3 What is the kinetic energy of a 1000 kg car travelling at 10 m/s? [Higher Tier only]

P2 3.3 Momentum

KEY POINTS

1 Momentum (kg m/s) = mass (kg) $\times$ velocity (m/s)
2 Momentum is conserved whenever objects interact, as long as no external forces act on them.

GET IT RIGHT!

The units of momentum are kg m/s or N m.

All moving objects have momentum.

Momentum is calculated using the equation:

$$\text{momentum} = \text{mass} \times \text{velocity}$$

The units of momentum are kilogram metre/second, kg m/s.

Whenever objects interact, the total momentum before the interaction is equal to the total momentum afterwards, provided no external forces act on them.

This is called 'conservation of momentum'.

Another way to say this is that the total change in momentum is zero.

The interaction could be a collision or an explosion. After a collision the objects may move off together, or they may move apart.

Key words: mass, velocity, momentum

A contact sport

CHECK YOURSELF

1 When do objects have momentum?
2 What are the units of momentum?
3 What is the momentum of a 1000 kg car travelling at 10 m/s?

P2 3.4 More on collisions and explosions

1 Momentum has size and direction.
2 When two objects push each other apart, they move apart with equal and opposite momentum.

GET IT RIGHT!

The total momentum before an explosion is usually zero. So the total momentum afterwards will be zero.

EXAM HINTS

In calculations it often helps to sketch a quick diagram to show what the objects are doing before and after the collision.

Like velocity, momentum has both size and direction. In calculations one direction must be defined as positive, so momentum in the opposite direction is negative.

When two objects are at rest their momentum is zero. In an explosion the objects move apart with equal and opposite momentum. One momentum is positive and the other negative, so the total momentum after the explosion is zero.

Firing a bullet from a gun is an example of an explosion. The bullet moves off with a momentum in one direction and the gun 'recoils' with equal momentum in the opposite direction.

Key words: direction, positive, negative, equal, opposite, explosion

CHECK YOURSELF

1 What must the total momentum after an explosion be equal to?

2 Two physics students on roller skates stand holding each other in the playground. They push each other away. What can you say about the momentum of each student?

3 One student has twice the mass of the other. What can you say about the velocity of each student?

P2 3.5 Changing momentum

1 The more time an impact takes, the less the force exerted.

2 Force (N) =

$$\frac{\text{change of momentum (kg m/s)}}{\text{time taken for change (s)}}$$

[Higher Tier only]

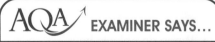

EXAMINER SAYS...

Force multiplied by time taken to change momentum is called the impulse.

When a force acts on an object that is moving, or able to move, its momentum changes. The equation that describes this is:

$$\text{force} = \frac{\text{change in momentum}}{\text{time taken for the change}}$$

HIGHER

For a particular change in momentum, the longer the time taken for the change, the smaller the force.

In a collision, the momentum of an object often becomes zero during the impact – the object comes to rest. If the impact time is short, the forces on the object are large. As the impact time increases, the forces become less.

This idea is the basis of a number of safety features in cars. Crumple zones in cars are designed to fold in a collision. This increases the impact time and so reduces the force on the car and the people in it.

Air bags work in a similar way. The driver's head changes momentum slowly when it hits an airbag. So the force on the head is less than if it changes momentum quickly by hitting the steering wheel.

Key words: impact time, change in momentum, safety features

BUMP UP YOUR GRADE

Make sure that you can explain how crumple zones and air bags reduce the forces acting by increasing the time taken to change the momentum of a car.

A crash test

CHECK YOURSELF

1 What is the equation that relates force, change in momentum and time? [Higher Tier only]

2 A car has a momentum of 30 000 kg m/s. Calculate the force needed to stop the car in 12 s. [Higher Tier only]

3 Why is a gymnast less likely to injure herself if she lands on a thick foam mat than if she lands on a hard floor?

P2 3   End of chapter questions

1 What are the units of energy transferred?

2 What force must be applied to an object if 2800 J of work are done moving it 7 m?

3 What is the equation that relates mass, kinetic energy and speed? [Higher Tier only]

4 What is the mass of a car that has 4000 J of kinetic energy when moving at 10 m/s? [Higher Tier only]

5 What is the momentum of a 2000 kg truck when it is travelling at 20 m/s?

6 A trolley of mass 0.2 kg moving at 1.5 m/s to the right collides with a stationary trolley of mass 0.3 kg. After the collision they move off together. Calculate the velocity of the trolleys after the collision.

7 What is the total momentum after a collision equal to?

8 Why might you calculate a value for velocity that is negative?

9 A hammer of mass 1 kg hits a nail when travelling at 0.5 m/s. It takes 0.2 s to come to rest. What is the force applied to the nail? [Higher Tier only]

10 A force of 2000 N is needed to bring a car to rest in 8 s. What is the initial momentum of the car? [Higher Tier only]

1 The diagram shows the forces acting on a flying helicopter.

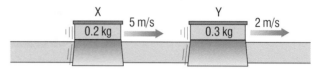

(a) What is the name of:
  (i) force X
  (ii) force Y?
  (2 marks)

(b) Describe the motion of the helicopter, if the force X is equal in size to the lift force and force Y is equal in size to the thrust force. (2 marks)

(c) Which force must decrease to make the helicopter:
  (i) decrease its height
  (ii) decrease its forward speed? (2 marks)

2 The diagram shows a man pushing a box up a slope with a force of 200 N.

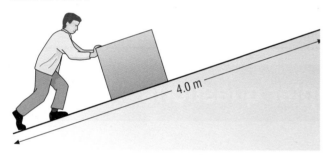

(a) The mass of the box is 50 kg. The gravitational field strength is 10 N/kg. Calculate the weight of the box. (2 marks)

(b) Calculate the work done pushing the box up the slope. (2 marks)

3 The drawing shows a ball being hit by a cricket bat.

(a) The ball has a mass of 0.16 kg. The bat is in contact with the ball for 0.02 s. The force exerted by the bat on the ball is 200 N.
  (i) Write down the equation that links change in momentum, force and time. (1 mark)
  (ii) Calculate the velocity of the ball, in m/s, away from the bat. (4 marks)

(b) A bowler bowls the ball with a speed of 25 m/s.
  Calculate, in joules, the kinetic energy of the ball. (3 marks)
  [Higher]

4 A student is investigating the conservation of momentum. She is using a horizontal air track and two gliders.

(a) What is the *conservation of momentum*? (2 marks)

(b) In what types of event does the conservation of momentum apply? (2 marks)

(c) The diagram shows the air track and the two gliders, X and Y.

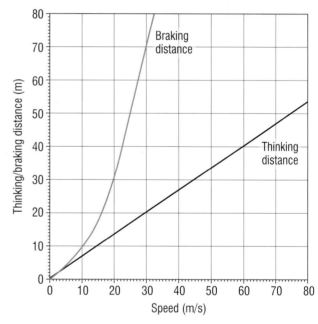

The mass of X is 0.2 kg and its velocity is 5 m/s. It collides with Y, which has a mass of 0.3 kg and is moving in the same direction at 2 m/s. The two gliders join together.

Calculate the speed of the gliders after the collision. (4 marks)

5 The graph below shows how the thinking distance and braking distance for a car increase with speed.

(a) Comment on how thinking distance and braking distance increase with speed. (2 marks)

(b) What is the total stopping distance at 25 m/s? (4 marks)

(c) What is the effect on thinking distance and braking distance of:
  (i) driving on a wet road surface? (2 marks)
  (ii) driving when very tired? (2 marks)

 Test & Assessment Interactive quizzes, answers and hints online!

The answer is worth 4 of the 6 marks available.

The responses worth a mark are underlined in red.

We can improve the answer in several ways:

Many cars are fitted with seat belts and driver airbags. Use your understanding of momentum to explain why airbags and seat belts reduce injuries to the driver in the event of a high speed crash. *(6 marks)*

Change in momentum = force × time
The airbag increases the time taken for the momentum change.
So force on the driver is less
So there is less chance of serious injury

The question asks about airbags and seatbelts, but seatbelts are not mentioned.

To gain the other 2 marks state that a driver going at a high speed will have a large momentum, and so a large change in momentum when they stop.

The answer is worth 5 of the 8 marks available for this Higher Tier question.

The responses worth a mark are underlined in red.

We can improve the answer in several ways:

A car travels on a straight, level road. The graph shows how the speed of the car changes with time.

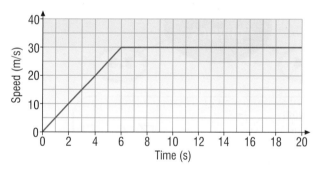

(a) Calculate the acceleration of the car during the first 6 seconds. *(4 marks)*

(b) Calculate the distance travelled by the car during the first 20 seconds of the journey. *(4 marks)*

(a) acceleration = slope

$$acceleration = \frac{30 - 0}{6}$$

acceleration = 5 m/s

(b) distance travelled = area under graph
distance travelled = (1/2 × 6 × 30) + (30 × 20)
distance travelled = 690 m

An incorrect unit is given in the answer. The units for speed (m/s) and acceleration (m/s²) have been mixed up, so no final mark.

The final answer is wrong but the unit is correct, so one of the final 2 marks is given.

The length of the rectangle is incorrectly written as 20, not 14.

P2 | Additional physics (Chapters 4–7)

Checklist

This spider diagram shows the topics in the second half of the unit. You can copy it out and add your notes and questions around it, or cross off each section when you feel confident you know it for your exams.

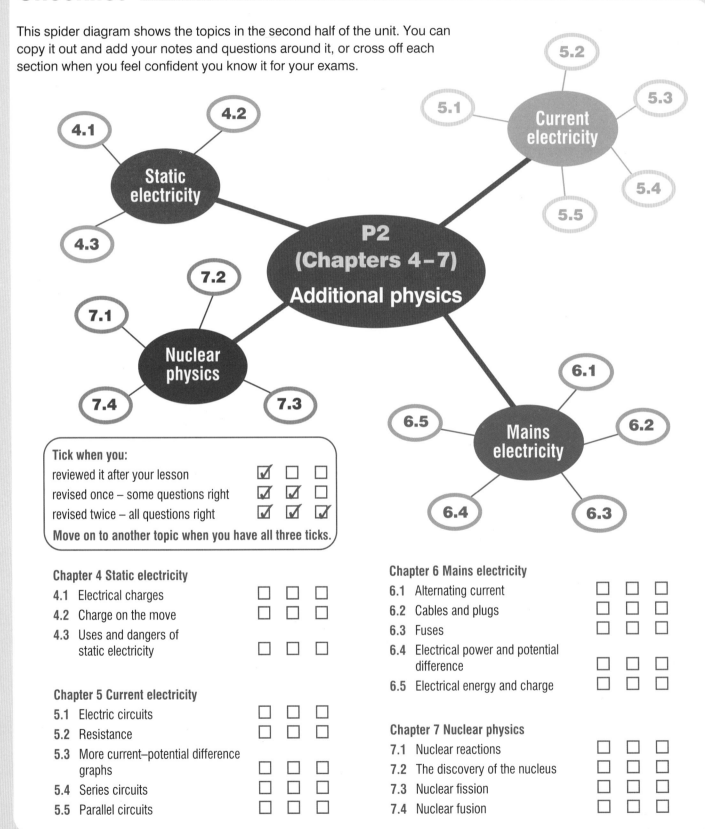

Tick when you:

reviewed it after your lesson	☑	☐	☐
revised once – some questions right	☑	☑	☐
revised twice – all questions right	☑	☑	☑

Move on to another topic when you have all three ticks.

Chapter 4 Static electricity

4.1	Electrical charges	☐	☐	☐
4.2	Charge on the move	☐	☐	☐
4.3	Uses and dangers of static electricity	☐	☐	☐

Chapter 5 Current electricity

5.1	Electric circuits	☐	☐	☐
5.2	Resistance	☐	☐	☐
5.3	More current–potential difference graphs	☐	☐	☐
5.4	Series circuits	☐	☐	☐
5.5	Parallel circuits	☐	☐	☐

Chapter 6 Mains electricity

6.1	Alternating current	☐	☐	☐
6.2	Cables and plugs	☐	☐	☐
6.3	Fuses	☐	☐	☐
6.4	Electrical power and potential difference	☐	☐	☐
6.5	Electrical energy and charge	☐	☐	☐

Chapter 7 Nuclear physics

7.1	Nuclear reactions	☐	☐	☐
7.2	The discovery of the nucleus	☐	☐	☐
7.3	Nuclear fission	☐	☐	☐
7.4	Nuclear fusion	☐	☐	☐

What are you expected to know?

Chapter 4 Static electricity (See students' book pages 160–167)

- When insulating materials are rubbed against each other they may become charged.
- Opposite charges attract, like charges repel.
- Electrostatic charge can be both useful and dangerous.

Chapter 5 Current electricity (See students' book pages 170–181)

- Common circuit symbols.
- Current–potential difference graphs for resistor, lamp and diode.
- Ohm's law: Potential difference = current × resistance
- Using Ohm's law with series and parallel circuits.
- Variation in the resistance of thermistors and light dependent resistors.

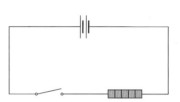

Chapter 6 Mains electricity

(See students' book pages 184–195)

- Correct wiring of a three-pin plug.
- How fuses are used.
- Charge = current × time
- Power = current × potential difference
- Energy transformed = potential difference × charge

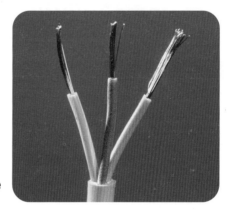

Chapter 7 Nuclear physics (See students' book pages 198–207)

- How the Rutherford–Marsden scattering experiment lead to the nuclear model of the atom.
- Relative charges and masses of protons, neutrons and electrons.
- The meaning of atomic number, mass number and isotope.
- Origins of background radiation.
- The effect of alpha and beta decay on nuclei.
- The process of fission and the meaning of a chain reaction.
- The process of fusion.

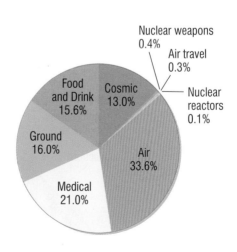

(1) How does an insulator become negatively charged?

(2) What will happen if two negatively charged objects are brought close to each other?

(3) What is meant by a 'conductor' of electricity?

(4) Why are metals good conductors?

(5) In an electrostatic smoke precipitator, how do the smoke particles become charged?

(6) What happens to the charged smoke particles?

---

students' book
page 160

P2 4.1   Electrical charges

KEY POINTS

1 Like charges repel; unlike charges attract.
2 Insulating materials that lose electrons when rubbed become positively charged.
3 Insulating materials that gain electrons when rubbed become negatively charged.

When two electrically insulating materials are rubbed together electrons are rubbed off one material and deposited on the other. Which way the electrons are transferred depends on the particular materials used.

Electrons have a negative charge so the material that has gained electrons becomes negatively charged. The one that has lost electrons is left with a positive charge.

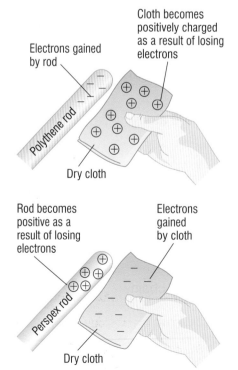

Cloth becomes positively charged as a result of losing electrons

Electrons gained by rod

Polythene rod

Dry cloth

Rod becomes positive as a result of losing electrons

Electrons gained by cloth

Perspex rod

Dry cloth

Charging by friction

GET IT RIGHT!

It is only ever electrons that move to produce positive and negative static charges on objects.

Two objects that have opposite electric charges are attracted to each other. Two objects that have the same electric charges repel each other. The bigger the distance between the objects, the weaker the force.

Key words: insulating, electrons, negative, positive, static, attract, repel

CHECK YOURSELF

1 What sort of charge does an electron have?

2 How does an insulator become positively charged?

3 What sort of force will there be between two negatively charged objects?

P2 4.2 Charge on the move

KEY POINTS

1 Electric current is the rate of flow of charge.
2 A metal object can only hold charge if it is isolated from the ground.
3 A metal object is earthed by connecting it to the ground.

GET IT RIGHT!

If a negatively charged, isolated conductor is earthed, electrons will flow from it to the earth until it is no longer charged. It the conductor is positively charged, electrons will flow from the earth to it.

When charge flows through a conductor there is a current in it. Electric current is the rate of flow of charge.

In a solid conductor, e.g. a metal wire, the charge carriers are electrons.

Metals are good conductors of electricity because they contain free, conduction electrons that are not confined to a single atom.

Insulators cannot conduct because all the electrons are held in atoms.

A conductor can only hold charge if it is isolated from the ground. Otherwise electrons will flow to or from the earth and discharge it.

The bigger the charge on an isolated object, the higher the potential difference between the object and the earth. If the potential difference becomes high enough, a spark may jump across the gap between the object and any earthed conductor brought near it.

HIGHER

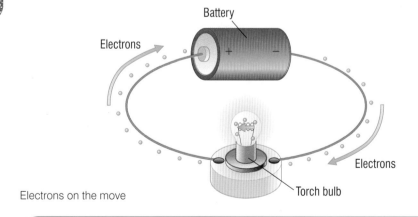

Electrons on the move

CHECK YOURSELF

1 Why can't insulators conduct?

2 How can a metal conductor be made to hold charge?

3 What is electric current?

KEY POINTS

1 A spark from a charged object can make powder grains or certain gases explode.
2 To eliminate static electricity:
   - use antistatic materials
   - earth metal pipes and objects.

EXAM HINTS

Make sure you understand these uses and dangers of static electricity.

In the exam you might be asked to explain one of them in detail.

In some situations electrostatic charge can be useful, and in some it can be dangerous.

Objects such as cars panels and bicycle frames are often painted with an electrostatic paint sprayer. The spray nozzle is connected to a positive terminal. As the paint droplets pass through it, they pick up a positive charge. The paint drops repel each other so they spread out to form a fine cloud. The item being painted is connected to a negative terminal so the positively charged droplets are attracted to it.

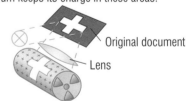

An electrostatic paint sprayer

In a photocopier, a copying plate is given a charge. An image of the page to be copied is projected onto the charged plate. Where light hits the plate the charge leaks away, leaving a pattern of the page. Black ink powder is attracted to the charged parts of the plate. This powder is transferred onto a piece of paper. The paper is heated so the powder melts and sticks to it, producing a copy of the original page.

1 Photocopiers with a photoconducting drum – drum positively charged until light falls on it.

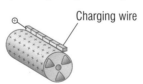

Charging wire

2 Light reflected off the paper onto the drum. The areas of black do not reflect so the drum keeps its charge in these areas.

Original document

Lens

3 The black toner sticks to the drum where it is still charged and is pressed onto paper.

Toner

4 The paper is finally heated to stick the toner to it permanently.

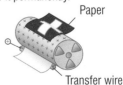

Paper

Transfer wire

Inside a photocopier

EXAM HINTS

Don't panic if you see a question in the exam about a use or danger of static electricity that you haven't seen before.

In these questions, you are expected to apply what you know about static electricity to an unfamiliar situation.

Electrostatic smoke precipitators are used in chimneys to attract dust and smoke particles so that they are not released into the air. The particles pass a charged grid and pick up charge. They are then attracted to plates on the chimney walls that have the opposite charge. The particles stick to the plates until they are shaken off and collected.

When powder or grain flows through a pipe, friction makes it pick up charge. This could lead to a spark igniting the powder, causing an explosion.

Ash and dust collect on plates

Grid of charged wires

Metal plates charged oppositely to the grid wires

Waste gases carrying ash and dust

An electrostatic precipitator

The filler pipes on road tankers that are used to pump fuel into storage tanks are earthed to prevent them becoming charged. A spark could cause an explosion of the fuel vapour.

Key words: electrostatic paint sprayer, photocopier, smoke precipitator

CHECK YOURSELF

1 Why are the filler pipes on fuel tankers earthed?

2 Why is the spray nozzle of an electrostatic paint sprayer connected to a positive terminal?

3 How does grain pick up charge as it flows through a pipe?

P2 4   End of chapter questions

1 Why do objects become positively or negatively charged only by the movement of negative charges?

2 Why do you sometimes get a shock from synthetic clothing when you take it off?

3 What is a lightning conductor made of?

4 How can an isolated, charged conductor be discharged?

5 Give three examples of devices that make use of static electricity.

6 A moving car may become charged. What would happen if you touched the door of the car?

1. What does a battery consist of?

2. What is the circuit symbol for a fuse?

3. What is the equation relating current, potential difference and resistance?

4. What is the unit of current?

5. What is the circuit symbol for a thermistor?

6. What happens to the resistance of a thermistor if its temperature increases?

7. How would you calculate the current in a series circuit?

8. Why is the current through each component in a series circuit the same?

9. Why is the potential difference (p.d.) across each component connected in parallel the same?

10. How would you calculate the total current in a parallel circuit?

---

students' book
page 170

P2 5.1 Electric circuits

KEY POINTS

1 Every component has its own agreed symbol.
2 A circuit diagram shows how components are connected together.
3 A battery consists of two or more cells connected together.

Every electrical component has an agreed symbol. Some of them are shown below:

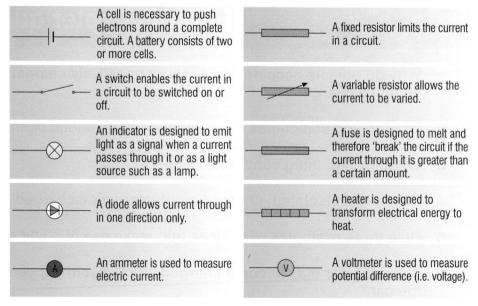

A cell is necessary to push electrons around a complete circuit. A battery consists of two or more cells.

A fixed resistor limits the current in a circuit.

A switch enables the current in a circuit to be switched on or off.

A variable resistor allows the current to be varied.

An indicator is designed to emit light as a signal when a current passes through it or as a light source such as a lamp.

A fuse is designed to melt and therefore 'break' the circuit if the current through it is greater than a certain amount.

A diode allows current through in one direction only.

A heater is designed to transform electrical energy to heat.

An ammeter is used to measure electric current.

A voltmeter is used to measure potential difference (i.e. voltage).

Components and symbols

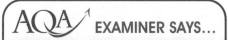 EXAMINER SAYS...

Make sure you can recognise and draw all of these circuit symbols.

Use a sharp pencil and make sure there are no gaps or odd ends of wire shown on your diagram.

GET IT RIGHT!

A battery consists of two or more cells connected together.

A circuit diagram uses the symbols to show how components are connected together to make a circuit.

Key words: circuit symbol

CHECK YOURSELF

1 What is the circuit symbol for a variable resistor?

2 What is the circuit symbol for a diode?

3 Draw a circuit diagram for a circuit containing a cell, a lamp, a resistor and a switch connected one after the other.

---

students' book page 172 P2 **5.2** **Resistance**

KEY POINTS

1 Resistance (ohms, Ω) =

$$\frac{\text{potential difference (volts, V)}}{\text{current (amperes, A)}}$$

2 The current through a resistor at constant temperature is directly proportional to the potential difference across the resistor.

GET IT RIGHT!

A graph with a straight line that passes through the origin shows that the two variables are directly proportional to each other.

AQA EXAMINER SAYS...

Remember that the resistor must stay at a constant temperature or the graph will not be a straight line.

Current–potential difference graphs are used to show how the current through a component varies with the potential difference across it.

The current is measured with an ammeter. Ammeters are always placed in series with the component. The unit of current is the ampere, A.

The potential difference is measured with a voltmeter. Voltmeters are always placed in parallel with the component. The unit of potential difference is the volt, V.

If the resistor is kept at a constant temperature the graph shows a straight line passing through the origin. This means the current is directly proportional to the potential difference across the resistor.

A current–p.d. graph for a wire at constant temperature

Potential difference and current are related by an equation called Ohms law:

potential difference = current × resistance.

Resistance is measured in ohms, Ω, and is the opposition to charge flowing through the resistor. If the resistor is kept at a constant temperature, the resistance stays constant. If the temperature of the resistor increases, the resistance increases. So the line on the current–p.d. graph is no longer straight.

Key words: current, potential difference, resistance

CHECK YOURSELF

1 What device is used to measure current?

2 What is the unit of resistance?

3 What do we mean by 'resistance'?

P2 5.3 More current–potential difference graphs

KEY POINTS

1 In a filament lamp, resistance increases with increase of the filament temperature.
2 In a diode, 'forward' resistance is low, 'reverse' resistance is high.
3 In a thermistor, resistance decreases if its temperature increases.
4 In an LDR, resistance decreases if the light intensity on it increases.

The current–potential difference graph for a filament lamp curves. So the current is not directly proportional to the potential difference.

The resistance of the filament increases as the current increases. This is because the resistance increases as the temperature increases.

Reversing the potential difference makes no difference to the shape of the curve.

The current through a diode flows in one direction only. In the reverse direction the diode has a very high resistance so the current is zero.

As the light falling on it gets brighter, the resistance of a light-dependent resistor (LDR) decreases.

As the temperature goes up, the resistance of a thermistor goes down.

A current–potential difference graph for a diode

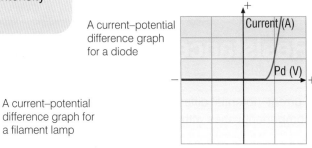

A current–potential difference graph for a filament lamp

Key words: filament lamp, diode, light dependent resistor, thermistor

CHECK YOURSELF

1 What happens to the resistance of a LDR if its surroundings become darker?
2 What effect does reversing the potential difference across a filament lamp have?
3 Explain the shape of a current–potential difference graph for a diode.

P2 5.4 Series circuits

KEY POINTS

For components in series:
1 The current is the same in each component.
2 The potential differences add to give the total potential difference.
3 The resistances add to give the total resistance.

GET IT RIGHT!

Remember that in a series circuit the current is the same everywhere in the circuit:

potential difference = current × resistance

This can be used to find the potential difference across each individual component if its resistance is known.

In a series circuit the components are connected one after another, so if there is a break anywhere in the circuit charge stops flowing. There is no choice of route for the charge as it flows around the circuit so the current through each component is the same.

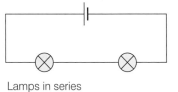

Lamps in series

The current depends on the potential difference (p.d.) of the supply and the total resistance of the circuit:

$$\text{current} = \frac{\text{p.d. of supply}}{\text{total resistance}}$$

The p.d. of the supply is shared between all the components in the circuit. So the p.d.s across individual components add up to give the p.d. of the supply.

The resistances of the individual components in series add up to give the total resistance of the circuit.

The bigger the resistance of a component, the bigger its share of the supply p.d.

Key words: series, supply

CHECK YOURSELF

1 What happens in a series circuit if one component stops working?
2 How could you find the total resistance in a series circuit?
3 A series circuit contains a variable resistor. If its resistance is increased, what happens to the p.d. across it?

P2 5.5 Parallel circuits

KEY POINTS

For components in parallel:
1 The potential difference is the same across each component.
2 The total current is the sum of the currents through each component.
3 The bigger the resistance of a component, the smaller its current is.

GET IT RIGHT!

In everyday life parallel circuits are much more useful than series circuits, as a break in one part of the circuit does not stop current flowing in the rest of the circuit.

In a parallel circuit each component is connected across the supply, so if there is a break in one part of the circuit charge can still flow in the other parts.

Each component is connected across the supply p.d., so the p.d. across each component is the same.

There are junctions in the circuit so different amounts of charge can flow through different components. The current through each component depends on its resistance. The bigger the resistance of a component, the smaller the current through it.

The total current through the whole circuit is equal to the sum of the currents through the separate components.

Key words: parallel

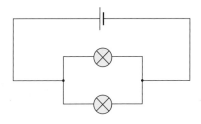

Lamps in parallel

EXAMINER SAYS…

Make sure that you understand the differences between series and parallel circuits.

CHECK YOURSELF

1 What happens in a parallel circuit if one component stops working?

2 In a parallel circuit what is the relationship between the supply p.d. and the p.d. across each parallel component?

3 How can you find the total current in a parallel circuit?

P2 5 End of chapter questions

1 What is a circuit diagram?

2 Draw a circuit diagram for a circuit containing a battery, a resistor, a variable resistor, a switch and an ammeter.

3 Where is an ammeter placed in a circuit?

4 What device is used to measure potential difference?

5 What is the circuit symbol for a light dependent resistor?

6 Why does the current–potential difference graph for a filament lamp curve?

7 A 2 Ω, a 6 Ω and a 10 Ω resistor are placed in series. What is their total resistance?

8 The total resistance in a series circuit is 24 Ω and the p.d. of the supply is 12 V. What is the current in the circuit?

9 A parallel circuit contains a variable resistor. If its resistance increases, what happens to the p.d. across it?

10 A parallel circuit contains a variable resistor. If its resistance increases, what happens to the current through it?

1. What is alternating current?

2. What is the frequency of the UK mains supply?

3. What colour is the live wire?

4. What is the outer cover of a three-pin plug made of?

5. What does a fuse contain?

6. What will happen if the live wire touches the metal case of an appliance?

7. What is the equation that relates power, energy and time?

8. What is the unit of electrical power?

9. What is the equation that relates charge, current and time?

10. What energy transformation takes place when charge flows through a resistor?

---

students' book page 184

P2 6.1 Alternating current

KEY POINTS

1. Alternating current repeatedly reverses its direction.
2. Mains electricity is an alternating current supply.
3. A mains circuit has a live wire which is alternately positive and negative every cycle and a neutral wire at zero volts. [Higher Tier only]

Cells and batteries supply current that passes round the circuit in one direction. This is called direct current, or d.c.

The current from the mains supply passes in one direction, then reverses and passes in the other direction. This is called alternating current, or a.c.

The frequency of the UK mains supply is 50 Hertz, which means it alternates direction 50 times each second. The 'voltage' of the mains is 230 V.

> The live wire of the mains supply alternates between a positive and a negative potential with respect to the neutral wire. The neutral wire stays at zero volts.
>
> The live wire alternates between +325 volts and −325 volts. In terms of electrical power, this is equivalent to a direct potential difference of 230 volts.

HIGHER

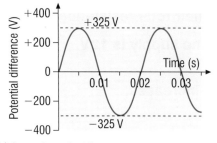

Mains p.d. against time

Key words: direct current, alternating current, live, neutral

EXAM HINTS

Make sure you can take readings from diagrams of an oscilloscope trace.

CHECK YOURSELF

1. What is direct current?
2. What is the potential of the neutral terminal? [Higher Tier only]
3. What is the potential difference of the mains supply?

P2 6.2 Cables and plugs

KEY POINTS

1 Cables consist of two or three insulated copper wires surrounded by an outer layer of flexible plastic material.
2 Sockets and plugs are made of stiff plastic materials, which enclose the electrical connections.
3 In a three-pin plug or a three-core cable:
   – The live wire is brown.
   – The neutral wire is blue.
   – The earth wire is yellow/green.
   – The earth wire is used to earth the metal case of a mains appliance.

Mains cable

EXAM HINTS

Make sure you can identify faults in the wiring of a three-pin plug.

Mains electricity can be dangerous unless used with care. Avoid hazards such as:

● Overlong or frayed cables.
● Using electricity near sources of heat or water.
● Overloading sockets with too many adaptors and plugs.

Most electrical appliances are connected to the mains supply using cable and a three-pin plug.

The outer cover of a three-pin plug is made of plastic or rubber, as these are good insulators. The pins of the plug are made of brass. Brass is a good conductor. It is also hard and will not rust or oxidise.

It is important that the cable grip is fastened tightly over the cable. There should be no bare wires showing inside the plug and the correct cable must be connected firmly to the terminal of the correct pin.

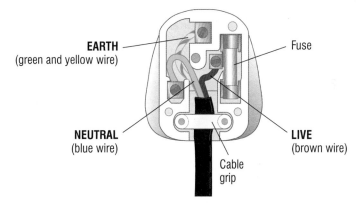

EARTH
(green and yellow wire)

Fuse

NEUTRAL
(blue wire)

LIVE
(brown wire)

Cable grip

Inside a three-pin plug

● The brown wire is connected to the live pin.
● The blue wire is connected to the neutral pin.
● The green-yellow wire (of a three-core cable) is connected to the earth pin. A two-core cable does not have an earth wire.

Appliances with metal cases must be earthed – the case is attached to the earth wire in the cable. Appliances with plastic cases do not need to be earthed. They are said to be double insulated and are connected to the supply with two-core cable containing just a live and a neutral wire.

Key words: insulator, three-pin plug, cable, live, neutral, earth

CHECK YOURSELF

1 Why must appliances with metal cases be earthed?

2 What colour is the neutral wire?

3 Why are the pins of the plug made of brass?

P2 6.3 Fuses

KEY POINTS

1 A fuse contains a thin wire that heats up and melts and cuts the current off if too much current passes through it.
2 A circuit breaker is an electromagnetic switch that opens (i.e. 'trips') and cuts the current off if too much current passes through it.

EXAM HINTS

Make sure you can explain how the earth wire and the fuse work together to protect an appliance.

Appliances with metal cases need to be earthed. Otherwise if a fault develops and the live wire touches the metal case the case becomes live and could give a shock to anyone that touches it.

If a fault develops in an earthed appliance a large current flows to earth and melts the fuse, disconnecting the supply.

A fuse must be put in the live wire so that if it melts it cuts off the current. The rating of the fuse should be slightly higher than the normal working current of the appliance. If it is much higher it will not melt soon enough. If it is not higher than the normal current it will melt as soon as the appliance is switched on.

A circuit breaker can be used in place of a fuse. This is an electromagnetic switch that opens and cuts off the supply if the current is bigger than a certain value.

Key words: fuse, circuit breaker

CHECK YOURSELF

1 What is a circuit breaker?
2 What happens if a fault develops in an earthed appliance?
3 Why do appliances with plastic cases not need to be earthed?

P2 6.4 Electrical power and potential difference

KEY POINTS

1 The power supplied to a device is the energy transferred to it each second.
2 Electrical power supplied (watts, W) = current (amperes, A) × potential difference (volts, V)

1650 – 1960 W
220 – 230 V ~
50 – 60 Hz

An electrical device transforms electrical energy to other forms and transfers energy from one place to another.

The rate at which it does this is called the power. Power can be calculated using:

$$\text{power (watts, W)} = \frac{\text{energy transformed (joules, J)}}{\text{time (seconds, s)}}$$

In an electric circuit it is more usual to measure the current through a device and the potential difference across it.

We can also use current and p.d. to calculate the power of a device:

power (watts, W) = current (amperes, A) × potential difference (volts, V)

Electrical appliances have their power rating shown on them. The p.d. of the mains supply is 230 V. So this equation can be used to calculate the normal current through an appliance and so work out the size of fuse to use.

Key words: power, energy

GET IT RIGHT!

Practice calculating the current in appliances of different powers so you can choose the correct fuse.

CHECK YOURSELF

1 What is the power of a mains appliance that takes a current of 10 A?
2 How much energy is transformed when a 3000 W appliance is used for 30 seconds?
3 What fuse should be used in a 500 W mains heater?

Power rating

P2 6.5 Electrical energy and charge

HIGHER

KEY POINTS

1 An electric current is the rate of flow of charge.
2 When charge flows through a resistor, electrical energy is transformed into heat energy.

GET IT RIGHT!

When charge flows through a resistor electrical energy is converted to heat. Since almost every component has electrical resistance, including connecting wires, when a charge flows in the circuit the components will heat up. This means that most electrical appliances have vents to keep them cool.

When charge flows through an appliance, electrical energy is transformed to other forms. In a resistor electrical energy is transformed into heat.

The amount of energy transformed can be calculated using the equation:

energy transformed = potential difference × charge
(joules, J)   (volts, V)   (coulombs, C)

When there is a current of one amp for one second the charge flowing is one coulomb. The equation relating charge, current and time is:

charge   =   current   ×   time
(coulombs, C)   (ampere, A)   (second, s)

Key words: energy, current, charge

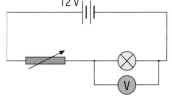

Energy transformations in a circuit

CHECK YOURSELF

1 What is the unit of charge? [Higher Tier only]

2 How much energy is transformed to heat when a charge of 200 C flows through a heater that has a potential difference across it of 230 V? [Higher Tier only]

3 How much charge flows past a particular point in a circuit when a current of 2 A flows for 2 minutes? [Higher Tier only]

P2 6   End of chapter questions

1 The UK mains supply has a frequency of 50 Hz. What does this mean?

2 What is the difference between d.c. and a.c.?

3 How should the cable grip on a plug be fixed?

4 What does the cover on an earth wire look like?

5 In a three-pin plug, which terminal is the fuse connected to?

6 Why must the fuse used in an appliance have a slightly higher rating than the normal working current?

7 What is the equation that relates power, current and potential difference?

8 What is the current through a 2300 W mains heater?

9 What is the equation that relates energy transformed, potential difference and charge? [Higher Tier only]

10 What is measured in coulombs? [Higher Tier only]

(1) What is the atomic number of a nucleus?

(2) What is the mass number of a nucleus?

(3) What was the plum pudding model of the atom? [Higher Tier only]

(4) What material were alpha particles fired at in Rutherford's experiment? [Higher Tier only]

(5) What is nuclear fission?

(6) Most uranium is uranium-238. What does the 238 tell you about the structure of the nucleus?

(7) What is nuclear fusion?

(8) Why do nuclei repel each other?

students' book
page 198

P2 7.1 Nuclear reactions

KEY POINTS

		Change in the nucleus	Particle emitted
1	α decay	The nucleus loses 2 protons and 2 neutrons	2 protons and 2 neutrons emitted as an α particle
2	β decay	A neutron in the nucleus changes into a proton	An electron is created in the nucleus and instantly emitted

GET IT RIGHT!

When a nucleus emits gamma radiation there is no change in the atomic number or the mass number, because a gamma ray is an electromagnetic wave which has no charge and no mass.

An atom has a nucleus, made up of protons and neutrons, surrounded by electrons.

The table below gives the relative masses and the relative electric charges of a proton, a neutron and an electron:

	Relative mass	Relative charge
proton	1	+1
neutron	1	0
electron	0.0005	−1

In an atom the number of protons is equal to the number of electrons, so the atom has no overall charge. If an atom loses or gains electrons it becomes charged and is called an 'ion'.

All atoms of a particular element have the same number of protons. Atoms of the same element that have different numbers of neutrons are called 'isotopes'.

The number of protons in an atom is called its 'atomic number'.

The total number of protons and neutrons in an atom is called its 'mass number'.

An alpha particle consists of two protons and two neutrons. When a nucleus emits an alpha particle the atomic number goes down by two and the mass number goes down by four.

For example, radium emits an alpha particle and becomes radon.

$$^{226}_{88}\text{Ra} \rightarrow \, ^{222}_{86}\text{Rn} + \, ^{4}_{2}\alpha$$

A beta particle is a high speed electron from the nucleus. It is emitted when a neutron in the nucleus changes to a proton and an electron. The proton stays in the nucleus so the atomic number goes up by one and the mass number is unchanged. The electron is instantly emitted.

For example, carbon-14 emits a beta particle when it becomes nitrogen:

$$^{14}_{6}\text{C} \rightarrow \, ^{14}_{7}\text{N} + \, ^{0}_{-1}\beta$$

Background radiation is the radiation that is around us all the time. It comes from many different sources such as cosmic rays, from rocks, or from nuclear power stations.

Key words: atomic number, mass number

EXAM HINTS

Make sure you can use nuclear equations to show how the atomic number and mass number change when alpha or beta particles are emitted.

CHECK YOURSELF

1 Where does background radiation come from?

2 What happens to the mass number of a nucleus when it emits a beta particle?

3 What happens to the atomic number of a nucleus when it emits an alpha particle?

---

students' book
page 200

P2 7.2 The discovery of the nucleus

HIGHER

KEY POINTS

1 Alpha particles in a beam are sometimes scattered through large angles when they are directed at a thin metal foil.

2 Rutherford used the measurements from alpha-scattering experiments to prove that an atom has a small positively charged central nucleus where most of the mass of the atom is located.

At one time scientists thought that atoms consisted of spheres of positive charge with electrons stuck into them, like plums in a pudding. So this became known as the 'plum pudding' model of the atom.

Rutherford, Geiger and Marsden devised an alpha particle scattering experiment, in which they fired alpha particles at thin gold foil.

Most of the alpha particles passed straight through the foil. This means that most of the atom is just empty space.

Some of the alpha particles were deflected through small angles, this suggests that the nucleus has a positive charge. A few rebound through very large angles. This suggests that the nucleus has a large mass and a very large positive charge.

Key words: alpha particle scattering

GET IT RIGHT!

The alpha particle has a positive charge. Because some of the alpha particles rebound, they must be repelled by another positive charge.

CHECK YOURSELF

1 What was Rutherford's alpha particle scattering experiment?

2 Why did most alpha particles pass straight through the foil in Rutherford's experiment?

3 What did the alpha particle scattering experiment suggest about the structure of the nucleus?

students' book page 202

KEY POINTS

1 Nuclear fission occurs when a neutron collides with and splits a uranium-235 nucleus or a plutonium-239 nucleus.
2 A chain reaction occurs when neutrons from the fission go on to cause further fission.
3 In a nuclear reactor one fission neutron per fission on average goes on to produce further fission.

BUMP UP YOUR GRADE

Make sure that you can draw a simple diagram to show a chain reaction.

Nuclear fission is the splitting of an atomic nucleus.

There are two fissionable isotopes in common use in nuclear reactors, uranium-235 and plutonium-239.

Naturally occurring uranium is mostly uranium-238, which is non-fissionable. Most nuclear reactors use 'enriched' uranium that contains 2–3% uranium-235.

For fission to occur the uranium-235 or plutonium-239 nucleus must absorb a neutron. The nucleus then splits into two smaller nuclei and 2 or 3 neutrons and energy is released. The energy released in such a nuclear process is much greater than the energy released in a chemical process such as burning.

The neutrons produced go on to produce further fissions, starting a chain reaction.

In a nuclear reactor the process is controlled, so one fission neutron per fission on average goes on to produce further fission.

Key words: fission, fissionable isotope, chain reaction

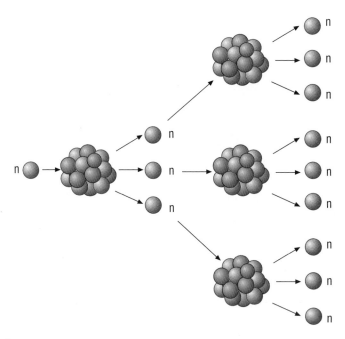

A chain reaction

CHECK YOURSELF

1 What is 'enriched' uranium?

2 What happens for fission to occur?

3 Which two fissionable isotopes are used in nuclear reactors?

P2 7.4 Nuclear fusion

KEY POINTS

1 Nuclear fusion occurs when two nuclei are forced close enough together so they form a single larger nucleus.
2 Energy is released when two light nuclei are fused together.
3 A fusion reactor needs to be at a very high temperature before nuclear fusion can take place.

AQA EXAMINER SAYS...

In an examination, students often confuse fission and fusion. Make sure that you can explain the difference between them.

Nuclear fusion is the joining of two atomic nuclei to form a single, larger nucleus.

During the process of fusion energy is released. Fusion is the process in which energy is released in stars.

There are enormous problems with producing energy from nuclear fusion in reactors. Nuclei approaching each other will repel one another due to their positive charge. To overcome this the nuclei must be heated to very high temperatures to give them enough energy to overcome the repulsion and fuse. Because of the enormously high temperatures involved the reaction cannot take place in a normal 'container', but has to be contained by a magnetic field.

Key words: fusion

A fusion reaction

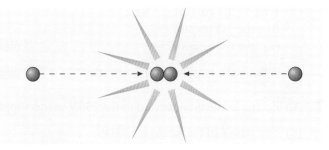

CHECK YOURSELF

1 By what process is energy released in stars?
2 How can nuclei be made to come close enough to fuse?
3 How are nuclei contained in a fusion reactor?

P2 7 End of chapter questions

1 **What happens to the mass number of a nucleus when it emits an alpha particle?**

2 **What happens to the atomic number of a nucleus when it emits a beta particle?**

3 **What particles were used in Rutherford's scattering experiment? [Higher Tier only]**

4 **What does most of an atom consist of?**

5 **What is a 'fissionable isotope'?**

6 **What is a 'chain reaction'?**

7 **Where does most nuclear fusion occur?**

8 **Why do nuclei not normally become close enough to fuse?**

1 The diagram shows a parallel circuit.

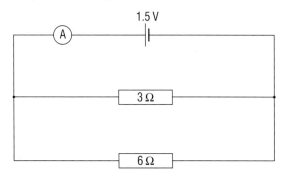

(a) What is the potential difference across:
(i) the $3\,\Omega$ resistor
(ii) the $6\,\Omega$ resistor? (2 marks)

(b) What is the current through:
(i) the $3\,\Omega$ resistor
(ii) the $6\,\Omega$ resistor? (3 marks)

(c) What is the reading on the ammeter? (2 marks)

2 An electric kettle is connected to the 230 V mains supply.

(a) The power of the kettle is 2300 W.
What is the current through the kettle? (3 marks)

(b) 3 A, 10 A and 13 A fuses are available. What size fuse should be used in the plug for the kettle? (1 mark)

(c) When the kettle is used to boil a litre of water, 2400 C of charge flows through the kettle. How long, in minutes, does it take to boil a litre of water?
(4 marks)
[Higher]

3 The diagram represents three different atoms X, Y and Z.

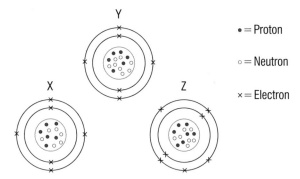

• = Proton
○ = Neutron
× = Electron

(a) Which two of the atoms have the same mass number? (1 mark)

(b) Which two of the atoms have the same atomic number? (1 mark)

(c) Which two of the atoms are from the same element? (1 mark)

(d) What must happen to atom Z to turn it into a positively charged ion? (1 mark)

(e) Atom Y decays by the emission of a beta particle. Explain what happens to the numbers of protons and neutrons in its nucleus. (2 marks)

4 A student is investigating how the potential difference across a resistor varies with the current through it. The student connects a cell, a resistor and a variable resistor in series.

(a) Draw a circuit diagram for the circuit. (4 marks)

(b) (i) How should the student connect an ammeter in the circuit to measure the current in the resistor?
(ii) How should the student connect a voltmeter in the circuit to measure the potential difference across the resistor? (2 marks)

(c) Sketch a graph to show how the potential difference across the resistor (x-axis) varies with the current through it (y-axis). (2 marks)

5 The process of nuclear fission may lead to a chain reaction.

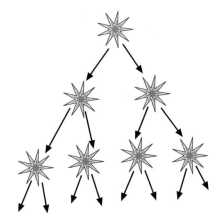

Explain the process of nuclear fission and how it can lead to a chain reaction. (6 marks)

There are 8 possible marks in total for this Higher Tier question.

This answer is worth 6 marks.

The responses worth a mark are underlined in red.

We can improve the answer in several ways:

The Rutherford and Marsden scattering experiment led to the 'plum pudding' model of the atom being replaced by the nuclear model.

(a) What is the 'plum pudding' model of the atom? *(2 marks)*

(b) The diagram shows the apparatus used for the scattering experiment to investigate the structure of the atom.

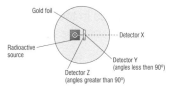

Gold foil
Detector X
Radioactive source
Detector Y (angles less then 90°)
Detector Z (angles greater than 90°)

(i) What is the charge on the particles emitted by the radioactive source? *(1 mark)*

(ii) The experiment took place in a vacuum. Suggest why. *(1 mark)*

(iii) Explain why detector X detected the most particles. *(2 marks)*

(iv) Some particles were scattered through very large angles and were detected by detector Z.
What does this suggest about the structure of a gold nucleus?
(2 marks)

This does not score a mark. A vacuum means that there are no gas molecules to get in the way and change the path of the alpha particles.

Stating 'this means that most of the atom is empty space' would gain the second mark.

(a) The plum pudding model says that the atom is a blob of positive matter with electrons stuck in like plums in a pudding.

(b) (i) Alpha particles are positively charged.

(ii) Easier to detect the particles.

(iii) Most particles go straight through and are not deflected.

(iv) The nucleus has a large positive charge and a large mass.

There are 8 marks available for this question.

This answer is worth 6 marks.

The responses worth a mark are underlined in red.

We can improve the answer in several ways:

The diagram below shows a circuit with a d.c. supply.

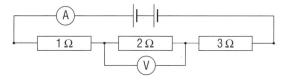

A
1 Ω   2 Ω   3 Ω
V

The potential difference of each cell is 1.5 V.

(a) What is the potential difference supplied by the battery? *(1 mark)*

(b) What is the total resistance in the circuit? *(2 marks)*

(c) Calculate the reading on the ammeter. *(3 marks)*

(d) Calculate the reading on the voltmeter. *(2 marks)*

The formula is correct and gets a mark. However, there are two cells so the supply p.d. of 1.5 V put into the equation is wrong. This makes the final answer wrong so the last 2 marks are lost.

The incorrect value from part (c) is put into the equation, but everything else is correct, so both marks are gained. Also, the working is shown at each stage.

(a) $1.5 V + 1.5 V = 3 V$

(b) Total resistance $= 1 \Omega + 2 \Omega + 3 \Omega$
Total resistance $= 6 \Omega$

(c) Current = supply p.d./total resistance
Current $= 1.5 V / 6 \Omega$
Current $= 0.25 A$

(d) p.d. = current × resistance
p.d. $= 0.25 A \times 2 \Omega$
p.d. $= 0.5 V$

P3 | Further physics (Chapters 1–2)

Checklist

This spider diagram shows the topics in the unit. You can copy it out and add your notes and questions around it, or cross off each section when you feel confident you know it for your exams.

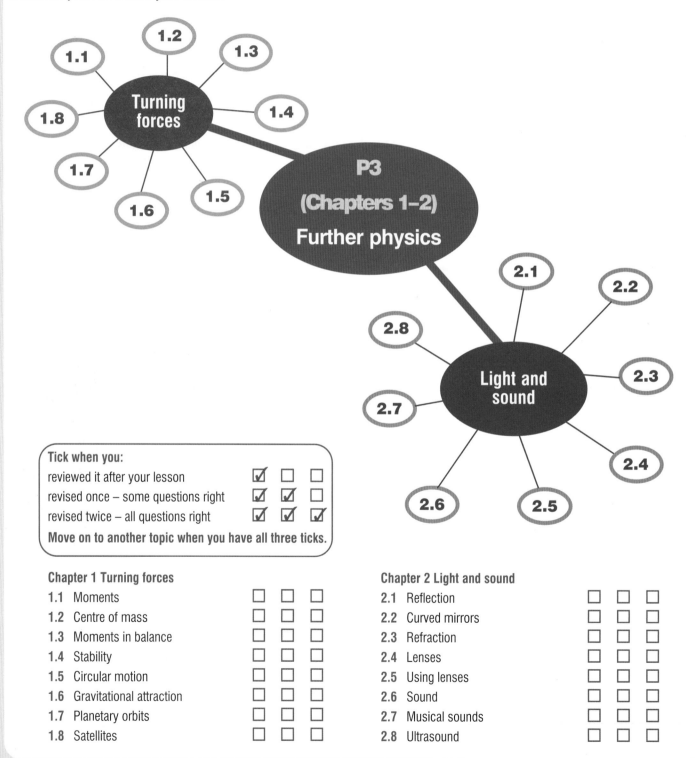

Tick when you:

reviewed it after your lesson	☑	☐	☐
revised once – some questions right	☑	☑	☐
revised twice – all questions right	☑	☑	☑

Move on to another topic when you have all three ticks.

Chapter 1 Turning forces

1.1	Moments	☐	☐	☐
1.2	Centre of mass	☐	☐	☐
1.3	Moments in balance	☐	☐	☐
1.4	Stability	☐	☐	☐
1.5	Circular motion	☐	☐	☐
1.6	Gravitational attraction	☐	☐	☐
1.7	Planetary orbits	☐	☐	☐
1.8	Satellites	☐	☐	☐

Chapter 2 Light and sound

2.1	Reflection	☐	☐	☐
2.2	Curved mirrors	☐	☐	☐
2.3	Refraction	☐	☐	☐
2.4	Lenses	☐	☐	☐
2.5	Using lenses	☐	☐	☐
2.6	Sound	☐	☐	☐
2.7	Musical sounds	☐	☐	☐
2.8	Ultrasound	☐	☐	☐

What are you expected to know?

Chapter 1 Turning forces (See students' book pages 214–231)

- The centre of mass of an object is the point where its mass can be thought to be concentrated.

- How to find the centre of mass of a thin sheet of a material.

- The moment of a force is the force × perpendicular distance from the line of action of the force to the axis of rotation (the pivot).

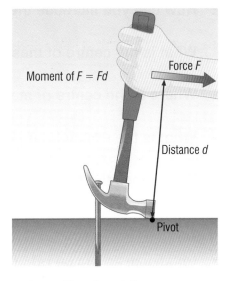

Moment of $F = Fd$

Force F

Distance d

Pivot

- If a body is not turning, the clockwise moment is equal to the anticlockwise moment. [Higher Tier only]

- The factors that affect the stability of a body. [Higher Tier only]

- Objects moving in a circle continuously accelerate towards the centre of the circle.

- The resultant force causing this acceleration is called the 'centripetal force'. This force always acts towards the centre of the circle.

- All bodies attract each other with a force called 'gravity'.

- Gravitational force provides the centripetal force that keeps planets and satellites in orbit.

- Communications satellites are usually put into geostationary orbits.

- Monitoring satellites are usually put into low polar orbits.

Chapter 2 Light and sound (See students' book pages 234–251)

- The angle of incidence is equal to the angle of reflection.

- The nature of images produced by plane, convex and concave mirrors.

- The nature of images produced by converging and diverging lenses.

- Sound is caused by mechanical vibrations, and travels as a wave.

- The pitch of a note increases as frequency increases.

- The loudness of a note increases as the amplitude increases.

- Ultrasound waves have a higher frequency than the limit of human hearing.

- Ultrasound waves are used in medicine and industry.

(1) What is the turning effect of a force called?

(2) How can you increase the turning effect of a particular force?

(3) What is the centre of mass of an object?

(4) Where is the centre of mass of a symmetrical body?

(5) What is the Principle of Moments? [Higher Tier only]

(6) When does the Principle of Moments apply to an object? [Higher Tier only]

(7) What factors affect the stability of an object? [Higher Tier only]

(8) When will an object topple? [Higher Tier only]

(9) What is centripetal acceleration?

(10) In which direction does a centripetal force act?

(11) What is the force of gravity?

(12) What happens to the force of gravity as bodies move further apart?

(13) What type of orbit do planets have around the Sun?

(14) What provides the centripetal force for planets in orbit?

(15) What is a communications satellite?

(16) What sort of orbit does a communications satellite have?

---

students' book
page 214 ## P3 1.1   Moments

KEY POINT

The moment of a force F about a pivot is $F \times d$, where d is the perpendicular distance from the pivot to the line of action of the force.

The turning effect of a force is called its 'moment'.

$$\text{moment of a force} = \text{force} \times \text{perpendicular distance from the pivot}$$
$$(\text{N m}) \qquad\qquad (\text{N}) \qquad\qquad \text{to the line of action of the force}$$
$$(\text{m})$$

GET IT RIGHT!

Notice that the equation uses the term 'perpendicular distance'. This means the shortest distance from the line that the force acts along.

EXAM HINTS

In examination questions, moments can be applied to lots of different situations such as:

- opening a door or a can of paint,
- moving something heavy in a wheel barrow or
- using a crowbar or a spanner

The idea is always the same, for any particular force make the distance to the pivot bigger to make the moment bigger.

To increase the moment:

- either the force must increase
- or the distance to the pivot must increase.

It is easier to undo a wheel nut by pushing on the end of a long spanner than a short one. That's because the long spanner increases the distance between the line of action of the force and the pivot.

Key words: moment, force, pivot

A turning effect

CHECK YOURSELF

1 A door opens when you apply a force of 20 N at right angles to it, 0.6 m from the hinge. What is the moment of the force?

2 What force would be needed to open the door if it were applied 0.3 m from the hinge?

3 Why is it easier to move a big rock with a crowbar than with your hands?

---

students' book page 216 P3 1.2 **Centre of mass**

KEY POINTS

1 The centre of mass of an object is the point where its mass may be thought to be concentrated.

2 When a suspended object is in equilibrium, its centre of mass is directly beneath the point of suspension.

3 The centre of mass of a symmetrical object is along the axis of symmetry.

AQA EXAMINER SAYS...

Make sure that you can describe the experiment to find the centre of mass of a thin sheet of a material, including sketching a labelled diagram.

Although any object is made up of many particles, its mass can be thought of as being concentrated at one single point. This point is called the 'centre of mass'.

Any body that is freely suspended will come to rest with its centre of mass directly below the point of suspension.

You can find the centre of mass of a thin sheet of a material as follows:

- Suspend the thin sheet from a pin held in a clamp stand. Because it is freely suspended, it is able to turn.

- When it comes to rest hang a plumbline from the same pin.

- Mark the position of the plumbline against the sheet.

- Hang the sheet with the pin at another point and repeat the procedure.

- The centre of mass is where the lines that marked the position of the plumbline cross.

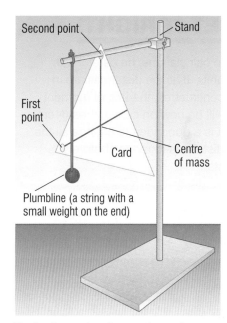

Finding the centre of mass of a card

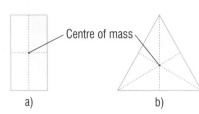

Centre of mass

a)   b)

Symmetrical objects

The position of the centre of mass depends on the shape of the object, and sometimes lies outside the object.

For a symmetrical object, its centre of mass is along the axis of symmetry. If the object has more than one axis of symmetry, the centre of mass is where the axes of symmetry meet.

Key words: centre of mass, plumbline, symmetrical, axis of symmetry

CHECK YOURSELF

1 Mark with a cross the position of the centre of mass of each of the shapes below.

(a)   (b)   (c)

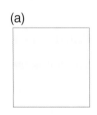  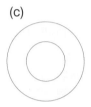

HIGHER

KEY POINTS

For an object in equilibrium: the sum of the anticlockwise moments about any point is equal to the sum of the clockwise moments about that point.

GET IT RIGHT!

In calculations, be careful with units. If all the distances are in centimetres, the unit of moment will be newton centimetres (N cm). It is a good idea to change all the distances to metres before you start.

 EXAMINER SAYS...

Be sure to add up all the clockwise moments and all the anticlockwise moments. Do not miss any out.

If an object is in equilibrium it is balanced, not turning. We can take the moments about *any* point and will find that the total clockwise moment and the total anticlockwise moment are equal.

There are lots of everyday examples of the Principle of Moments, such as seesaws and balance scales.

Key words: equilibrium, clockwise moment, anticlockwise moment

The seesaw

CHECK YOURSELF

1 Sam sits 2 m from the centre of a seesaw. Alex weighs twice as much as Sam. How far from the centre must he sit to balance the seesaw?

2 Joe weighs 600 N and sits 1.5 m from the centre of a seesaw. Sebastian sits 2 m from the centre of the seesaw to balance it. How much does he weigh?

3 If someone sits in the centre of the seesaw, the moment about the pivot is zero. Why?

P3 1.4 Stability

HIGHER

KEY POINTS

1 The stability of an object is increased by making its base as wide as possible and its centre of mass as low as possible.
2 An object will tend to topple over if the line of action of its weight is outside its base.

EXAM HINTS

Stability is important in the design of lots of different objects, and examination questions could include any of them. These could include double-decker buses, racing cars, high chairs, prams and filing cabinets. The stability of any object is increased by making its centre of mass lower and its base wider.

The line of action of the weight of an object acts through its centre of mass.

If the line of action of the weight lies outside the base of an object, there will be a moment and the object will tend to topple over.

The wider the base of an object, and the lower its centre of mass, the further it has to tilt before the line of action of the weight moves outside the base. So the stability of an object is increased by making its base wider and its centre of mass lower.

Key words: stability, topple, base, centre of mass

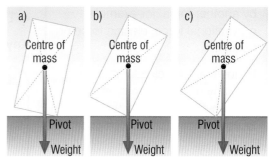
Tilting and toppling. a) Tilted, b) at balance, c) toppled over.

CHECK YOURSELF

1 Why does hanging heavy bags from the handle of a pushchair make it more likely to topple over?
2 Why do ten-pin bowling pins have a narrow base and a high centre of gravity?
3 Which of the glasses shown below would be the most stable?

A   B   C

P3 1.5 Circular motion

KEY POINTS

For an object moving in a circle at constant speed:
● the object accelerates continuously towards the centre of the circle
● the centripetal force on it increases as:
  1 The mass of the object increases.
  2 The speed of the object increases.
  3 The radius of the circle decreases.

When an object moves in a circle it is continuously changing direction, so it is continuously changing velocity. In other words, it is accelerating. This acceleration is called the 'centripetal acceleration'.

An object only accelerates when a resultant force acts on it. This force is called the 'centripetal force' and always acts towards the centre of the circle.

If the centripetal force stops acting, the object will continue to move in a straight line at a tangent to the circle.

Whirling an object round

BUMP UP YOUR GRADE

Centripetal force is not a force in its own right. It is always provided by another force, for example gravitational force, electric force or tension.

In questions on circular motion, you may need to identify the force that provides the centripetal force.

The centripetal force needed to make a body perform circular motion increases as:

- The mass of the body increases.
- The speed of the body increases.
- The radius of the circle decreases.

Key words: circle, centripetal acceleration, centripetal force, radius

CHECK YOURSELF

1 Why must an object moving in a circle be accelerating?

2 A student is whirling a conker around, on a piece of string, in a horizontal circle. What force provides the centripetal force?

3 What will happen to the conker if the string breaks?

KEY POINTS

The force of gravity between two objects:
- is an attractive force,
- is bigger, the greater the mass of each object,
- is smaller, the greater the distance between the two objects.

- Each body in the Universe attracts every other body with a force called 'gravity'.

- Often this force is too small to notice, but the greater the mass of the body the bigger the force.

- For objects with very large masses, like stars and planets, the force is very big and has a noticeable effect over large distances.

- As the distance between two bodies increases, the gravitational force between them decreases.

- The force of gravity from the Earth acting on an object is called the object's 'weight'.

Object   Object

Force of gravitational attraction

Gravitational attraction

GET IT RIGHT!

Each body in the Universe attracts every other body, but unless the bodies have big masses we don't usually notice the force.

If you drop something it always falls downwards. That's because the gravitational force attracts it towards the centre of the Earth.

Key words: gravity, force, attracts, mass, distance

CHECK YOURSELF

1 What happens to the force of gravity between two objects when they move closer together?

2 Why would you weigh less on the Moon than you do on the Earth?

3 Why don't you notice the gravitational force between yourself and other people?

AQA EXAMINER SAYS...

Remember that gravitational forces are always attractive.

Planetary orbits

KEY POINTS

1 To stay in orbit at a particular distance, a small body must move at a particular speed around a larger body.
2 The larger an orbit is, the longer the orbiting body takes to go round the orbit.

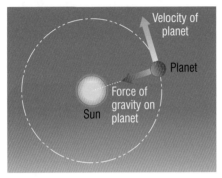

A circular orbit

EXAM HINTS

You do not need to know the distances from the Sun and times of orbit of the planets. However, you may be given a table of such information in the exam and be asked to use it with your knowledge of orbits to answer questions.

The planets of the Solar System orbit the Sun. The orbits are slightly squashed circles (ellipses) with the Sun quite close to the centre.

The centripetal force that keeps a planet moving in its orbit is provided by the gravitational force between the Sun and the planet.

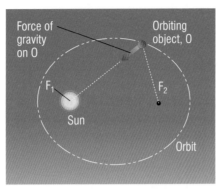

An elliptical orbit
1. The orbit is defined by the two 'foci' F_1 and F_2, (such that the sum of the distances OF_1 and OF_2 is constant).
2. The Sun is at one focus of the ellipse.

To stay in orbit at a particular distance, a planet (or any other body) must move at a particular speed around the Sun.

The further away from the Sun:

● The less the speed of the planet as it moves around the Sun.

● The longer the planet takes to make a complete orbit. For example, Mercury (the closest planet) takes 88 days to complete 1 orbit. On the other hand, Pluto (the furthest planet) takes 248 years to complete 1 orbit.

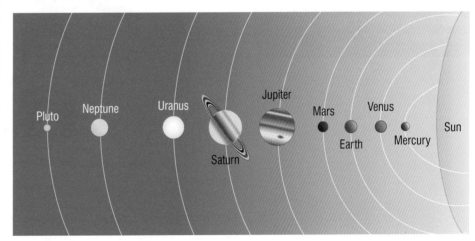

Planetary orbits (not to scale)

Key words: planets, Sun, orbit, gravitational force, centripetal force

CHECK YOURSELF

Mercury is the closest planet to the Sun:

1 What does this tell you about the time taken for it to complete one orbit?

2 What does this tell you about the speed at which Mercury moves in its orbit?

3 If there were no gravitational force, Mercury would not orbit the Sun. Why not?

KEY POINTS

1 A satellite in a geostationary orbit has a period of 24 hours and stays at the same position directly above the Earth's equator.
2 Geostationary orbits are usually used for communication satellites.
3 Monitoring satellites are usually in low polar orbits.

GET IT RIGHT!

All geostationary satellites orbit above the Equator.

AQA EXAMINER SAYS...

Make sure you can explain the differences between geostationary and polar orbits.

The Moon is a natural satellite. There are many different types of artificial satellite. They can be used for:

- monitoring climate and the weather
- spying
- space research
- navigation
- communication.

Communications satellites are placed in a geostationary orbit. This means that they stay above the same place on the Earth's surface. They do this by having an orbit above the Equator that takes 24 hours to complete. They are used for receiving and sending telephone and TV signals. The height of the orbit is about 36 000 kilometres above the Earth.

A satellite in orbit

Monitoring satellites are put into low polar orbits. They have an orbit that passes over both poles while the Earth rotates beneath them. Each orbit takes only two or three hours, and the whole Earth can be monitored each day. The height of the orbit is much lower than the height of a geostationary orbit.

Key words: satellites, communications, monitoring, geostationary

CHECK YOURSELF

1 How long does it take to complete one geostationary orbit?

2 Why does a TV satellite need to stay over the same point on the Earth's surface?

3 What is the Earth's natural satellite?

1 What do we mean by the 'moment of a force'?

2 What factors affect the moment of a force?

3 If an object is suspended, where will its centre of mass be when it comes to rest?

4 A ruler is balanced on a knife-edge. Where must its centre of mass be?

5 If a seesaw balances with a child sat at each end, what do you know about the weight of the children? [Higher Tier only]

6 If a child sits 1 m from the centre of a seesaw, where should his friend, who is twice as heavy, sit to balance it? [Higher Tier only]

7 Why does a racing car need a low centre of mass? [Higher Tier only]

8 Why does an object topple if the line of action of its weight is outside its base? [Higher Tier only]

9 What is the direction of a centripetal acceleration?

10 What factors affect the size of the centripetal force needed to make a body perform circular motion?

11 What sort of objects experience gravitational forces?

12 What factors affect the size of the gravitational force between two bodies?

13 What is an orbit?

14 How does the time taken for a planet to orbit the Sun change with its distance from the Sun?

15 What is a monitoring satellite?

16 What sort of orbit does a monitoring satellite have?

1. What is an angle of incidence?

2. What is an angle of reflection?

3. What does a concave mirror look like?

4. What is the principal focus (or focal point) of a concave mirror?

5. What is refraction?

6. What happens to the speed of light when it crosses from air into glass?

7. What does a converging lens look like?

8. What sort of image is formed by a diverging lens?

9. What type of lens is used in a camera?

10. What type of lens is used as a magnifying glass?

11. What is a reflected sound wave called?

12. Why can't sound waves be heard in space?

13. What does the pitch of a sound depend on?

14. What happens to a sound as the amplitude of the wave increases?

15. What is ultrasound?

16. How is ultrasound used in medicine?

P3 2.1 Reflection

KEY POINTS

1 The 'normal' at a point on a mirror is perpendicular to the mirror.
2 For a light ray reflected by a mirror:

   the angle of incidence = the angle of reflection

GET IT RIGHT!

The angle of incidence and the angle of reflection are always measured between the ray and the normal.

AQA EXAMINER SAYS...

Be sure that you understand the difference between a real image and a virtual image.

The image seen in a mirror is due to the reflection of light.

The diagram below shows how an image of a point is formed by a plane (flat) mirror.

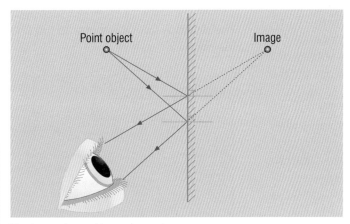

The image formed by a plane mirror

- The incident ray is the ray that goes towards the mirror.
- The reflected ray is the one coming away from the mirror.
- We draw a line, called the 'normal', at right angles (perpendicular) to the mirror at the point where the incident ray hits the mirror.
- The angle of incidence is the angle between the incident ray and the normal.
- The angle of reflection is the angle between the reflected ray and the normal.
- For *any* reflected ray the angle of incidence is equal to the angle of reflection.

We can see objects because light reflects off them:

- When light reflects from an uneven surface, such as a page of this book, it reflects at lots of different angles. This gives a diffuse reflection.
- When light reflects from an even surface, such as a mirror, it all reflects at the same angle. This gives a regular reflection.

The image in a plane mirror is:
- the same size as the object
- upright
- the same distance behind the mirror as the object is in front
- virtual.

Real or virtual image

- A real image is one that can be formed on a screen, because the rays of light that produce the image actually pass through it.
- A virtual image cannot be formed on a screen, because the rays of light that produce the image only appear to pass through it.

Key words: normal, incidence, reflection, mirror, image, real, virtual

CHECK YOURSELF

1 What size is the image in a plane mirror?

2 If you stand 1.5 m in front of a plane mirror, where will your image be?

3 What type of image is produced by a plane mirror?

KEY POINTS

1 The principal focus (or focal point) of a concave mirror is the point where parallel rays are focused to by the mirror.
2 A concave mirror forms:
  • a real image if the object is beyond the focal point of the mirror,
  • a virtual image if the object is between the mirror and the focal point.
3 A convex mirror always forms a virtual image of an object.

BUMP UP YOUR GRADE

Make sure that you can draw ray diagrams to find the position of the image in a plane, concave or convex mirror. Ray diagrams should be drawn accurately and neatly with a sharp pencil. Remember to put arrows on the rays to show their direction.

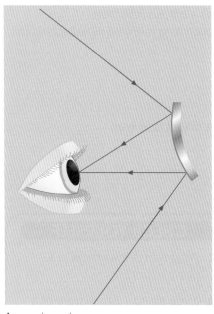

A rear-view mirror

Concave mirror

Light rays from the same point on a very distant object are effectively parallel when they reach the mirror. They are brought to focus at a point called the principal focus (or focal point) of the mirror. A real image of the object is formed here.

The distance from the mirror to the principal focus is called the focal length of the mirror.

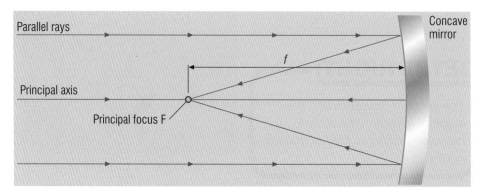

The focal length of a concave mirror

For an object placed beyond the principal focus of the mirror a real, inverted image is produced. The position and size of the image depends on the distance from the object to the mirror.

The magnification of the image can be calculated using the equation:

$$\text{magnification} = \frac{\text{image height}}{\text{object height}}$$

For an object placed between the principal focus and the mirror a virtual, upright image is produced. The image is behind the mirror and larger than the object.

Convex mirror

A convex mirror always produces a virtual, upright image that is smaller than the object and behind the mirror.

Convex mirrors are used as rear-view mirrors in cars. They produce a wider field of view than a plane mirror.

Key words: concave, convex, upright, inverted, magnification

CHECK YOURSELF

1 A concave mirror produces an image that is 7 mm tall from an object that is 5 mm tall. What is the magnification?

2 Which types of mirror always produce virtual images?

3 What is the principal focus (or focal point) of a curved mirror?

P3 2.3 Refraction

KEY POINTS

KEY POINTS

1 Refraction of light is the change of direction of a light ray when it crosses a boundary between two transparent substances.
2 If the speed is reduced, refraction is towards the normal (e.g. air to glass).
3 If the speed is increased, refraction is away from the normal (e.g. glass to air).

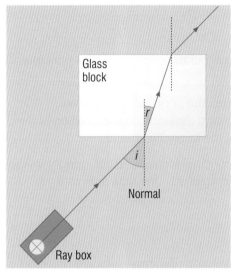

Refraction of light

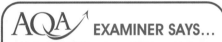

EXAMINER SAYS...

Do not confuse reflection and refraction.

GET IT RIGHT!

A ray of light travelling along a normal is not refracted.

Refraction is a property of all waves, including light and sound. It happens because waves change speed when they cross a boundary. The wavelength of the waves also changes, but the frequency stays the same.

The change in speed of the waves causes a change in direction, unless the waves are travelling along a normal.

Refraction of light means that a light ray changes direction when the light crosses a boundary between two substances, such as air and glass or air and water.

When light enters a more dense substance, such as going from air to glass, it slows down and the ray bends towards the normal.

When light enters a less dense substance, such as going from glass to air, it speeds up and the ray bends away from the normal.

Different colours of light have different wavelengths, and are refracted by slightly different amounts. When a ray of white light is shone onto a triangular glass prism we can see this because a spectrum is produced. This is called 'dispersion'.

● Blue light is refracted the most.

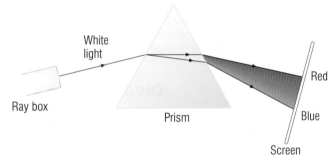

Dispersion of light by a prism

● Red light is refracted the least.

Key words: refraction, boundary, prism, spectrum, dispersion

CHECK YOURSELF

1 What is the spreading of white light into different colours by a prism called?
2 Which colour of light is refracted the least?
3 Why does refraction take place?

P3 2.4 Lenses

KEY POINTS

1 A real image is formed by a converging lens if the object is further away than its principal focus (focal point).
2 A virtual image is formed by a diverging lens, and by a converging lens if the object is nearer than the principal focus.

EXAMINER SAYS...

Don't confuse 'converging' and 'diverging'.

Converging lens

Parallel rays of light that pass through a converging (convex) lens are refracted so that they converge to a point. This point is called the principal focus (focal point). The distance from the centre of the lens to the principal focus is the focal length.

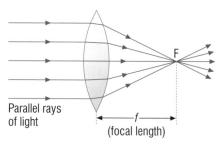

Parallel rays of light

f (focal length)

Converging lens

Because light can pass through the lens in either direction, there is a principal focus on either side of the lens.

If the object is further away from the lens than the principal focus, a real, inverted image is formed. The size of the image depends on the position of the object. The nearer the object is to the lens, the larger the image.

If the object is nearer to the lens than the principal focus, a virtual, upright image is formed behind the object. The image is magnified – the lens acts as a magnifying glass.

Diverging lens

Parallel rays of light that pass through a diverging (concave) lens are refracted so that they diverge away from a point. This point is called the principal focus. The distance from the centre of the lens to the principal focus is the focal length.

Because light can pass through the lens in either direction there is a principal focus on either side of the lens.

The image produced by a diverging lens is always virtual.

Key words: converging, diverging, focal point, focal length

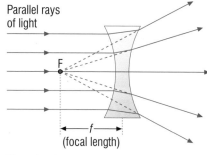

Parallel rays of light

f (focal length)

Diverging lens

CHECK YOURSELF

1 Where must an object be placed for a converging lens to produce a virtual image?

2 What is the focal length of a lens?

3 Which way up is the image when a converging lens is used as a magnifying glass?

KEY POINTS

1 A camera contains a converging lens that is used to form a real image of an object.
2 A magnifying glass is a converging lens that is used to form a virtual image of an object.

GET IT RIGHT!

Make sure you know the basic design for a camera.

EXAM HINTS

You might be asked to draw a ray diagram in the exam, so make sure that you practise. Only two of the construction rays are needed to find the image, but if you have time it is worth drawing all three to be sure that you have the correct position.

We can draw ray diagrams to find the image that different lenses produce with objects in different positions.

The line through the centre of the lens and at right angles to it is called the 'principal axis'. This should be included in the diagram.

Ray diagrams use three 'construction' rays from a single point on the object to locate the corresponding point on the image:

- A ray parallel to the principal axis is refracted through the principal focus.
- A ray through the centre of the lens travels straight on.
- A ray through the principal focus is refracted parallel to the principal axis.

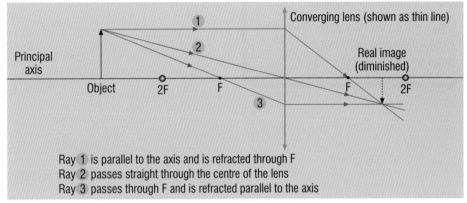

Ray 1 is parallel to the axis and is refracted through F
Ray 2 passes straight through the centre of the lens
Ray 3 passes through F and is refracted parallel to the axis

Formation of a real image by a converging lens

A camera uses a converging lens to form a real image of an object on a film or an array of pixels.

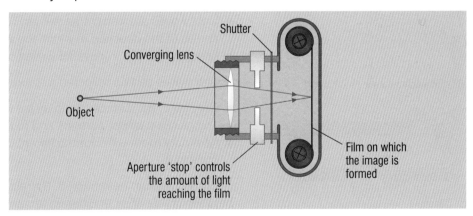

The camera

A magnifying glass uses a converging lens to form a virtual, magnified image of an object, on the same side of the lens as the object.

Key words: ray diagram, principal axis, construction rays, camera, magnifying glass

CHECK YOURSELF

1 What are 'construction' rays?

2 What type of image is formed in a camera?

3 Why can't a diverging lens be used in a camera?

P3 2.6 Sound

Sound waves:
1 can travel through liquids and gases and in solids,
2 cannot travel in a vacuum,
3 are longitudinal waves,
4 can be reflected (echoes) and refracted.

GET IT RIGHT!

Make sure you can describe the similarities and differences between light and sound.

Sound is caused by mechanical vibrations in a substance, and travels as a wave.

Sound can travel through liquids, solids and gases, but not through a vacuum. This can be tested by listening to a ringing bell in a 'bell jar'. As the air is pumped out of the jar, the ringing sound fades away.

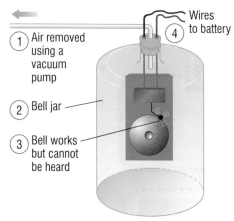

A sound test

Sound waves generally travel fastest in solids and slowest in gases.

The range of frequencies that can be heard by the human ear is from 20 Hz to 20 000 Hz. The ability to hear the higher frequencies declines with age.

- Sound waves are longitudinal waves. The direction of the vibrations is the same as the direction in which the wave travels.
- Light waves are transverse waves. The direction of the vibrations is at right angles to the direction in which the wave travels.
- Longitudinal and transverse waves can be demonstrated with a slinky spring.

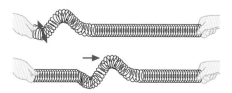

(a) Longitudinal waves

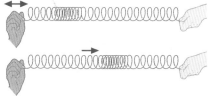

(b) Transverse waves

Sound waves can be reflected to produce echoes:

- Only hard, flat surfaces such as flat walls and floors reflect sound.
- Things like carpets, curtains and furniture absorb sounds.
- An empty room will sound different once carpets, curtains and furniture are put into it.

Sound waves can be refracted. Refraction takes place at the boundaries between layers of air at different temperatures.

Key words: longitudinal waves, transverse waves, frequency, reflection, refraction

CHECK YOURSELF

1 What is a longitudinal wave?

2 What is a transverse wave?

3 What range of frequencies can be heard by the human ear?

KEY POINTS

1 The loudness of a note depends on the amplitude of the sound waves.

2 The pitch of a note depends on the frequency of the sound waves.

AQA EXAMINER SAYS...

Make sure you understand the meanings of 'frequency' and 'amplitude'.

EXAM HINTS

Practise sketching waveforms, e.g. sketch a wave with twice the frequency and half the amplitude of the original.

- The bigger the amplitude of a wave, the more energy it carries and the louder the sound.
- The higher the frequency of the wave, the higher the pitch of the sound.

Differences in waveforms can be shown on an oscilloscope.

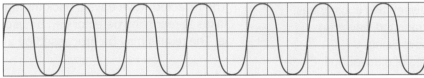

A) Loud and high-pitched

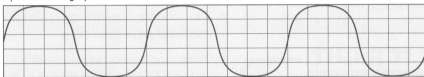

B) Loud and low-pitched

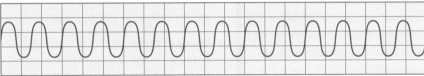

C) Quiet and high-pitched (higher pitch than A)

Investigating sounds

The quality of a note depends on the waveform.

Tuning forks and signal generators produce 'pure' waveforms.

A musical instrument produces a waveform that is a mixture of different frequencies. Instruments have different sounds because they produce different waveforms.

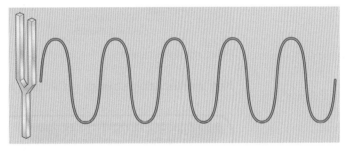

Tuning fork waves (a 'pure' waveform)

Flute wave pattern

Key words: loudness, amplitude, pitch, frequency, note, waveform

CHECK YOURSELF

1 Why do different instruments sound different when they play the same note?

2 What happens to the frequency of a note as the pitch increases?

3 How does a musical instrument create sound?

students' book
page 248

KEY POINTS

1 Ultrasound waves have a frequency above 20 000 kHz.
2 Ultrasound waves are partly reflected at a boundary between two substances.
3 Ultrasound waves are non-ionising.
4 Uses of ultrasound include cleaning devices, flaw detectors and medical scanners.

GET IT RIGHT!

In the time between a transmitter sending out a pulse of ultrasound and it returning to a detector it has travelled from the transmitter to a boundary and back, i.e. twice the distance to the boundary.

The human ear can detect sound waves with frequencies between 20 Hz and 20 000 Hz. Sound waves of a higher frequency than this are called 'ultrasound' (ultrasonic waves).

Electronic systems can be used to produce ultrasound waves. When a wave meets a boundary between two different materials, part of the wave is reflected. The reflected wave travels back through the material to a detector. The time it takes to reach the detector can be used to calculate how far away the boundary is. The results may be processed by a computer to give an image.

Uses

● Ultrasound is non-ionising. This makes it safer to use than X-rays to produce images of an unborn baby.

● Ultrasound is used to clean delicate mechanisms, such as watches, jewellery and spectacles without having to dismantle them. The object to be cleaned is put in a tank of water and ultrasonic waves are passed through the tank. The waves dislodge dirt particles.

● Flaws in metal castings can be detected with ultrasound. A transmitter on the surface sends waves into the metal object. These are reflected back from boundaries, including cracks or flaws in the metal, to reach a detector on the surface next to the transmitter.

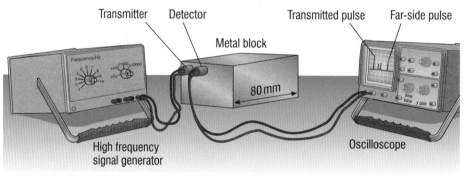

Detecting flaws in a metal

Key words: ultrasound, reflected, boundary, images

CHECK YOURSELF

1 Why can ultrasound be used to detect flaws in metal castings?

2 Why is it safer to use ultrasound than X-rays to produce an image of an unborn baby?

3 Why would ultrasound be used for cleaning a delicate mechanism like a watch?

1   What is a 'normal'?

2   What is the relationship between the angle of incidence and the angle of reflection?

3   What does a convex mirror look like?

4   What type of image is formed by a concave mirror?

5   What happens to the speed of light when it crosses from glass to air?

6   What happens to the direction of a ray of light as it crosses from glass to air?

7   What does a diverging lens look like?

8   What is the principal focus (focal point) of a lens?

9   Why is a converging lens used in a camera?

10   What type of image is formed by a magnifying glass?

11   Why can't sound waves travel through a vacuum?

12   What type of waves are sound waves?

13   What does the loudness of a sound depend on?

14   What happens to the pitch of a note as the frequency decreases?

15   Why can't we hear ultrasound?

16   Why is ultrasound used for looking at unborn babies?

1 The diagram shows two children sitting on a seesaw.

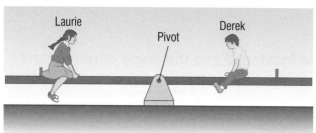

Laurie weighs 420 N and is sitting 2 m from the pivot. Derek is sitting 1.6 m from the pivot. The seesaw is balanced.

(a) Calculate the moment of Laurie's weight about the pivot. (2 marks)

(b) Calculate Derek's weight. (2 marks)

(c) Explain what would happen if the children were to sit one at each end of the seesaw. (2 marks)

[Higher]

2 (a) What is the 'centre of mass' of an object? (1 mark)

(b) The drawing shows a sheet of stiff card. Three small holes have been made in the card as shown.

Describe how you could use the following to find the centre of mass of the card:
- a clamp and stand
- a large pin
- a weight on a thin piece of string (a plumbline)
- a ruler and pen. (5 marks)

3 The diagram shows the position of an object in front of a plane mirror.

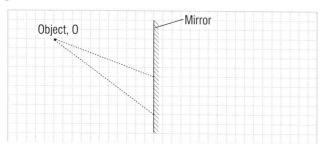

(a) On a piece of squared paper use a ruler to construct accurately the position of the image. (3 marks)

(b) The image produced is virtual. What is a 'virtual' image? (1 mark)

4 A camera uses a converging lens to produce a real image.

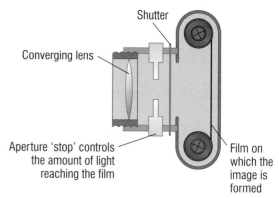

(a) (i) Explain why a real image must be produced in a camera. (2 marks)

(ii) How are the object and lens positioned to produce a real image that is smaller than the object? (1 mark)

(b) A converging lens can also be used as a magnifying glass.

(i) How are the object and lens positioned to produce a virtual image that is larger than the object? (1 mark)

(ii) Such a lens produces an image 4.0 cm tall of an object 1.6 cm tall. What is the magnification produced by the lens? (2 marks)

5 (a) What is ultrasound? (1 mark)

(b) Explain how ultrasound waves can be used to clean a delicate mechanism such as a watch. (3 marks)

(c) Explain how ultrasound waves can be used to detect flaws in a metal object. (4 marks)

 Test & Assessment Interactive quizzes, answers and hints online

The answer is worth 5 out of the 7 marks.

The responses worth a mark are underlined in red.

We can improve the answer in several ways:

This answer is incomplete so gains one of the marks. To get the other mark the student should have said that if the velocity of an object changes it is **accelerating**.

This is incomplete. It should be towards the centre **of the Earth** or the centre **of the orbit**.

The Moon moves in a circular orbit around the Earth. It moves with a constant speed but is continuously accelerating towards the Earth. The resultant force causing this acceleration is called the centripetal force.

(a) Explain how an object moving at a constant speed can be accelerating.

(2 marks)

(b) (i) Which force provides the centripetal force to keep the Moon in orbit? *(1 mark)*

(ii) In which direction does the centripetal force act? *(1 mark)*

(c) What would happen to the size of the centripetal force needed to keep the Moon in orbit if:

(i) the mass of the Moon were greater?

(ii) the radius of the orbit were greater ?

(iii) the speed of the Moon were greater? *(3 marks)*

(a) If an object changes its direction it also changes its velocity.

(b) (i) gravitational force

(ii) towards the centre

(c) (i) bigger

(ii) smaller

(iii) bigger

The student identifies the effect of all the factors correctly and gets all the available marks. The centripetal force increases as the mass and speed of the orbiting body increase, and as the radius of orbit decreases.

The answer is worth 4 out of the 6 marks.

The responses worth a mark are underlined in red.

We can improve the answer in several ways:

The amplitude is from the equilibrium position to the top of a crest or the bottom of a trough. So the student only scores 1 mark.

The answer gets 1 of the 2 marks. To get the second mark, the student should explain that in a vacuum there are no air particles to vibrate, so there is no sound.

The diagram shows the note produced by a musical instrument, on an oscilloscope screen.

(a) Draw an arrow on the diagram to show:

(i) the wavelength, *(1 mark)*

(ii) the amplitude of the wave. *(1 mark)*

(b) The diagram on the right shows how a different note played by the same instrument appears on the screen. How would the note shown in the second diagram sound different from the note shown in the first diagram? *(2 marks)*

(c) Explain why the sound from the musical instrument could not be heard through a vacuum. *(2 marks)*

(a) Wavelength   Amplitude   (b) louder
higher

(c) Sound is caused by the vibration of air particles.

P3 | Further physics (Chapters 3–4)

Checklist

This spider diagram shows the topics in the unit. You can copy it out and add your notes and questions around it, or cross off each section when you feel confident you know it for your exams.

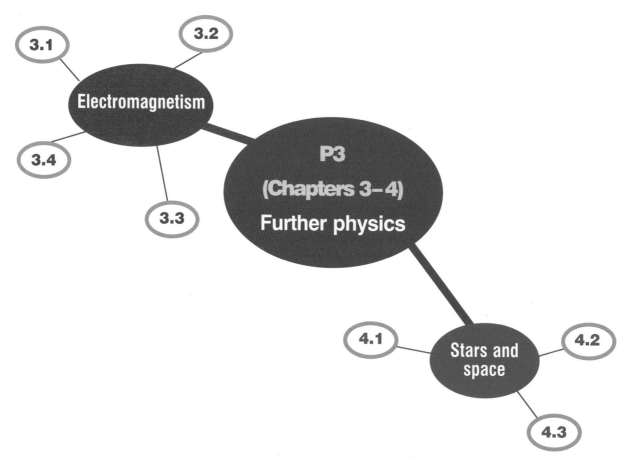

Tick when you:

reviewed it after your lesson	☑	☐	☐
revised once – some questions right	☑	☑	☐
revised twice – all questions right	☑	☑	☑

Move on to another topic when you have all three ticks.

Chapter 3 Electromagnetism

3.1	The motor effect	☐	☐	☐
3.2	Electromagnetic induction	☐	☐	☐
3.3	Transformers	☐	☐	☐
3.4	Transformers and the National Grid	☐	☐	☐

Chapter 4 Stars and space

4.1	Galaxies	☐	☐	☐
4.2	The life history of a star	☐	☐	☐
4.3	How the chemical elements formed	☐	☐	☐

What are you expected to know?

Chapter 3 Electromagnetism (See students' book pages 254–263)

- A conductor carrying an electric current in a magnetic field experiences a force.

- This effect is used in electric motors.

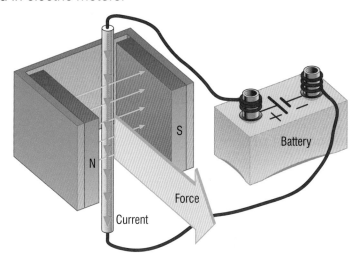

- If a conductor cuts through magnetic field lines, a p.d. (potential difference) is induced across the ends of the wire.

- Use of this effect in electric generators.

- The basic structure of a transformer.

- The use of step-up and step-down transformers in the National Grid.

Chapter 4 Stars and space (See students' book pages 266–273)

- The life cycle of stars.

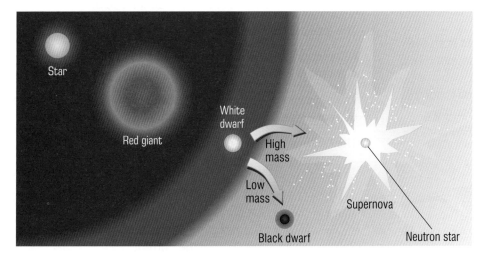

- Fusion processes in stars produce all naturally occurring elements. [Higher Tier only]

Pre Test: Electromagnetism

1. What happens when a wire carrying a current is placed in a magnetic field?

2. What happens to the wire in question 1 when the strength of the magnetic field is increased?

3. What happens when a wire cuts the lines of a magnetic field?

4. What happens when the wire is part of a complete circuit?

5. What does a transformer consist of?

6. What sort of supply must be used with a transformer?

7. What does a step-up transformer do?

8. What sort of transformer has more turns on the primary coil than the secondary coil?

---

students' book
page 254

P3 3.1 The motor effect

KEY POINTS

In the motor effect, the force:
1 is increased if the current or the strength of the magnetic field is increased,
2 is at right angles to both the direction of the magnetic field and to the wire,
3 is reversed if the direction of either the current or the magnetic field is reversed.

When we place a wire carrying an electric current in a magnetic field, it may experience a force. This is called the 'motor effect'.

The force is a maximum if the wire is at an angle of 90° to the magnetic field, and zero if the wire is parallel to the magnetic field.

The size of the force can be increased by:

- increasing the strength of the magnetic field
- increasing the size of the current.

The direction of the force on the wire is reversed if either the direction of the current or the direction of the magnetic field is reversed.

The motor effect is used in different devices.

The diagram shows a simple electric motor:

- The speed of the motor is increased by increasing the size of the current.
- The direction of the motor can be reversed by reversing the direction of the current.

When a current passes through the coil, the coil spins because:

- A force acts on each side of the coil due to the motor effect.
- The force on one side of the coil is in the opposite direction to the force on the other side.

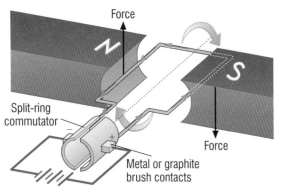

Force

Split-ring commutator

Force

Metal or graphite brush contacts

The electric motor

The split ring commutator reverses the direction of the current around the coil every half turn. Because the sides swap over each half-turn, the coil is always pushed in the same direction.

Key words: motor, magnetic field, current, direction, force, commutator

EXAMINER SAYS...

Make sure that you can explain the importance of the motor effect in devices such as a motor and loudspeaker.

CHECK YOURSELF

1 What happens to the direction of the force on a wire carrying a current if the direction of the current and the magnetic field are both reversed?

2 What does a commutator do?

3 The ends of the coil in a motor are parallel to the magnetic field. What is the size of the force on them?

---

students' book page 256

P3 3.2 Electromagnetic induction

KEY POINTS

1 When a wire cuts the magnetic field lines, a potential difference (p.d.) is induced in the wire.
2 If the wire is part of a complete circuit, the induced p.d. causes a current in the circuit.
3 The size of the current is increased if the wire moves faster, or if a stronger magnet is used.

BUMP UP YOUR GRADE

Make sure that you can identify the parts on a diagram of a dynamo and a generator and explain how each works.

GET IT RIGHT!

A p.d. is induced only when the wire or coil and the magnetic field move relative to each other.

If a wire cuts through magnetic field lines a potential difference (p.d.) is induced across the ends of the wire. If a magnet is moved into a coil of wire a p.d. is induced across the ends of the coil. If the wire or coil is part of a complete circuit a current passes through it.

If the direction of movement of the wire or coil is reversed or the polarity of the magnet is reversed, the direction of the induced p.d. is also reversed. A p.d. is only induced while there is movement.

The size of the induced p.d. is increased by increasing:

● the speed of movement
● the strength of the magnetic field
● the number of turns on the coil
● the area of the coil.

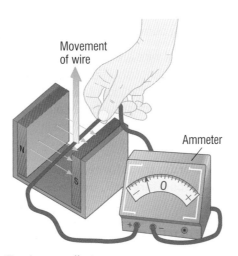

The dynamo effect

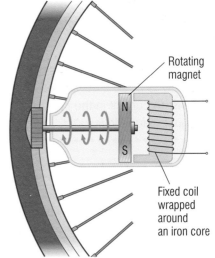

The dynamo

In a dynamo a magnet spins causing magnetic field lines to continually cut across the wires of a coil. This induces an alternating p.d.

In a generator, a coil is rotated inside a stationary magnetic field.

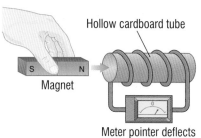

Hollow cardboard tube

Magnet

Meter pointer deflects when the magnet is pushed into the coil

Testing electromagnetic induction

The faster the coil rotates:

- the larger the peak value of the alternating current
- the greater the frequency of the alternating current.

Key words: magnetic field, induced, potential difference, dynamo, generator

CHECK YOURSELF

If a bar magnet is moved into a coil of wire a p.d. is induced across the ends of the wire:

1 What happens if the magnet is held at rest in the coil?

2 What happens if the magnet is pulled back out of the coil?

3 What happens if the magnet is held at rest and the coil is pulled off it?

---

students' book page 258 | **P3 3.3** | **Transformers**

KEY POINTS

1 A transformer consists of a primary coil and a secondary coil wrapped on the same iron core.

2 Transformers only work using alternating current.

EXAM HINTS

There is no current in the iron core, just magnetic flux.

GET IT RIGHT!

Transformers do not work with d.c. but only a.c. If a d.c. passes through the primary coil, a magnetic field is produced in the core but it would *not* be continually expanding and collapsing so no p.d. would be induced in the secondary coil.

A transformer consists of two coils of insulated wire, called the 'primary coil' and the 'secondary coil', which are wound onto the same iron core. When an alternating current passes through the primary coil, it produces an alternating magnetic field around the core which continually expands and collapses.

The alternating magnetic field lines pass through the secondary coil and induce an alternating potential difference across its ends. If the secondary coil is part of a complete circuit an alternating current is produced.

The coils of wire are insulated so that current does not short across either the iron core or adjacent turns of wire, but flows around the whole coil. The core is made of iron so it is easily magnetised.

Key words: transformer, primary coil, secondary coil, iron core

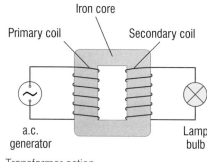

Transformer action

CHECK YOURSELF

1 Why are the coils on a transformer made from insulated wire?

2 Why is the core of a transformer made of iron not copper?

3 What happens if a 1.5 V cell is used as the supply for the primary coil?

P3 3.4 Transformers and the National Grid

KEY POINTS

1 Transformers are used to increase or decrease potential difference (p.d.).

2 Transformer equation:

$$\frac{\text{p.d. across primary, } V_P}{\text{p.d. across secondary, } V_S} = \frac{\text{number of turns on primary, } N_P}{\text{number of turns on secondary, } N_S}$$

[Higher Tier only]

EXAMINER SAYS...

Practise using the transformer equation.

- In a step-up transformer the p.d. across the secondary is greater than the p.d. across the primary.
- In a step-down transformer the p.d. across the secondary is less than the p.d. across the primary.

The p.d. across, and the number of turns on, the primary and secondary coils are related according to the equation:

$$\frac{\text{p.d. across primary, } V_P}{\text{p.d. across secondary, } V_S} = \frac{\text{number of turns on primary, } N_P}{\text{number of turns on secondary, } N_S}$$

The National Grid uses transformers to step-up the p.d. from power stations. This is because the higher the p.d. at which electrical energy is transmitted across the Grid the smaller the energy losses in the cables.

Step-down transformers are used to reduce the p.d. so that it is safe to be used by consumers.

Key words: transformer, step-up, step-down, National Grid

CHECK YOURSELF

1 Why is a transformer used to step-up the p.d. from a power station?

2 A transformer has 100 turns on the primary coil and 400 turns on the secondary coil. The p.d. across the primary coil is 2 V. What is the p.d. across the secondary coil?

3 The p.d. across the primary coil of a transformer is 2000 V and the p.d. across the secondary is 40 000 V. If there are 120 turns on the primary coil, how many turns are there on the secondary coil?

P3 3 End of chapter questions

1 **What is the 'motor effect'?**

2 **What direction is the force produced by the motor effect?**

3 **What happens if a magnet is moved into a coil of wire?**

4 **What happens if the speed with which the magnet moves is increased?**

5 **What is the core of a transformer made from?**

6 **What happens when an alternating current passes through the primary coil of a transformer?**

7 **What does a step-down transformer do?**

8 **Where are step-down transformers used in the National Grid?**

1. Which galaxy is our Sun part of?

2. How many galaxies does the Universe contain?

3. What is a black hole?

4. What process produces energy in stars?

5. What is nuclear fusion?

6. What was the only element present in the early Universe?

---

students' book
page 266

P3 4.1 Galaxies

KEY POINTS

1 As the Universe expanded, it cooled and uncharged atoms formed.
2 The force of gravity pulled matter into stars and galaxies.

GET IT RIGHT!

The distance between neighbouring stars is usually millions of times greater than the distance between planets in our Solar System. The distance between neighbouring galaxies is usually millions of times greater than the distance between stars within a galaxy. So the Universe is mostly empty space.

Most scientists believe that the Universe was created by a 'Big Bang' about 13 thousand million (13 billion) years ago. At first the Universe was a hot glowing ball of radiation. In the first few minutes the nuclei of the lightest elements formed. As it expanded, over millions of years, its temperature fell. Uncharged atoms were formed.

The force of gravity takes over

Eventually dust and gas were pulled together by gravitational attraction to form stars. The resulting intense heat started off nuclear fusion reactions in the stars, so they began to emit visible light and other radiation.

Gravitational force attracts more dust and matter around stars and this may have formed planets.

Very large groups of stars are called galaxies. Our Sun is one of the many billions of stars in the Milky Way galaxy. The Universe contains billions of galaxies.

Key words: Big Bang, gravitational attraction, stars, galaxies, Milky Way

CHECK YOURSELF

1 What is a galaxy?

2 What happened to the temperature of the Universe as it expanded?

3 How long did it take for the temperature to fall enough so that uncharged atoms were formed?

---

students' book page 268 **P3 4.2** # The life history of a star

KEY POINTS

1 Low mass star:
protostar → main sequence star → red giant → white dwarf → cold (black) dwarf
2 High mass star, after the white dwarf stage:
white dwarf → supernova → neutron star → black hole (if there is sufficient mass)

EXAM HINTS

Exam questions may ask you to put the stages of the life cycle of a star in the correct order, possibly by filling in details on a diagram, so make sure you learn the names of the stages and their order thoroughly.

Stars form from clouds of dust and gas.

Gravitational forces make the clouds become increasingly dense, forming a 'protostar'.

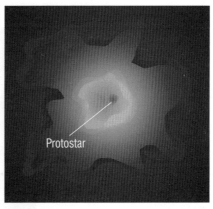
Protostar

Star birth

As a protostar becomes denser, it gets hotter. If it becomes hot enough, the nuclei of hydrogen atoms and other light elements start to fuse together. Energy is released in the process so the core gets hotter and brighter, and the star begins to shine.

Stars radiate energy because of hydrogen fusion in the core. This is the main stage in the life of a star. It can continue for billions of years until the star runs out of hydrogen nuclei to fuse together.

The inward force of gravity is balanced by the outward pressure of radiation from the core, so the star is stable. These forces stay in balance until most of the hydrogen nuclei in the core have been fused together.

During this stable period the star is called a 'main sequence star'. This period lasts for billions of years.

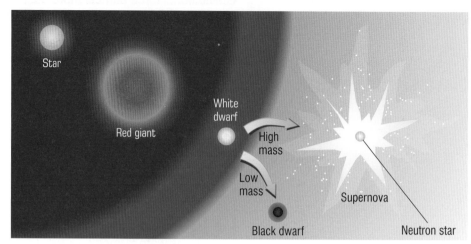

The end of a star

Eventually a star runs out of hydrogen nuclei. The star swells into a 'red giant'. It is red because the surface has cooled. What happens next in the life cycle of the star depends on its size. A star similar in size to our Sun will contract to form a 'white dwarf'. Eventually no more light is emitted and the star becomes a 'black (cold) dwarf'.

A star much larger than the Sun will continue to collapse and eventually explode in a 'supernova'. The outer layers are thrown out into space. The core is left as an extremely dense 'neutron star'.

If this is massive enough it becomes a 'black hole'. The gravitational field of a black hole is so strong that not even light can escape from it.

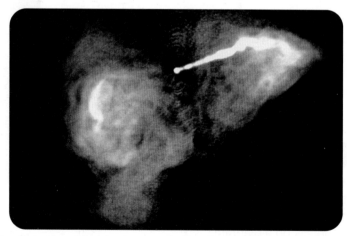

M87 is a galaxy that spins so fast that its centre is thought to contain a black hole that has a billion times more mass than the Sun

Key words: protostar, red giant, white dwarf, black dwarf, supernova, neutron star, black hole

CHECK YOURSELF

1 What is a neutron star?

2 Why is a black hole black?

3 Why are stars in the main sequence stable?

P3 4.3 How the chemical elements formed

HIGHER

KEY POINTS

1 Elements as heavy as iron are formed inside stars as a result of nuclear fusion.
2 Elements heavier than iron are formed in the final stages of a star's life in a supernova explosion.

Chemical elements are formed by fusion processes in stars. The nuclei of lighter elements fuse to form the nuclei of heavier elements. The process gives out large amounts of heat and light.

Elements heavier than iron are only formed in the final stages of the life of a big star. This is because the process requires the input of energy. All the elements get distributed through space by the supernova explosion.

The presence of the heavier elements in the Sun and inner planets is evidence that they were formed from debris scattered by a supernova.

Key words: fusion, elements

GET IT RIGHT!

In the process of fusion, light nuclei fuse to form heavier nuclei and energy is given out. For elements heavier than iron to be formed there must be an input of energy.

CHECK YOURSELF

1 What is the name of the process that produces the chemical elements?

2 When are elements heavier than iron formed?

3 What evidence is there that the Sun and inner planets formed from the remnants of a supernova?

P3 4 End of chapter questions

1 What force caused the formation of the stars and planets?

2 What material came together to form the stars and planets?

3 What determines what the life cycle of a star will be?

4 What is a 'supernova'?

5 How are elements heavier than hydrogen produced? [Higher Tier only]

6 How are the elements produced in stars spread through the Universe? [Higher Tier only]

1 Our Sun is one of the many millions of stars in the Milky Way galaxy.

(a) How do scientists believe that stars are formed?

(2 marks)

(b) A star goes through a life cycle.

Our Sun is in a stable period of its life. This period lasts for billions of years. Eventually the Sun will become a black dwarf.

(i) Explain, in terms of the forces acting on the Sun, why it is stable. (4 marks)

(ii) Explain what will happen to the Sun after the stable period and before it becomes a black dwarf. (4 marks)

2 (a) Copy and complete the sentences:

Chemical elements are formed by . . . processes in stars.

In this process the nuclei of lighter elements form the nuclei of . . . elements.

The process gives out large amounts of . . . .

The elements are distributed throughout space by a . . . explosion. (4 marks)

(b) The biggest stars may eventually become black holes. What is a black hole? (2 marks)

[Higher]

3 A student is doing an investigation with a magnet and a coil of wire attached to a meter.

When no current flows in the coil the meter looks like this.

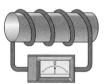

In her first experiment the student moves the magnet slowly into the coil. The position of the meter pointer is shown on the diagram below.

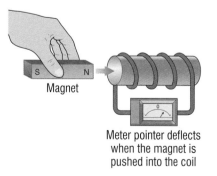

Magnet

Meter pointer deflects when the magnet is pushed into the coil

The student repeats the experiment several times, making the changes described below. Draw the position of the meter pointer in each case.

(a) She holds the magnet stationary inside the coil.

(b) She moves the magnet slowly back out of the top of the coil.

(c) She moves the magnet quickly into the coil.

(d) She moves the magnet slowly into the coil with the south pole first.

(e) She moves the magnet slowly into a coil with many more turns. (5 marks)

4 The diagram shows a simple transformer.

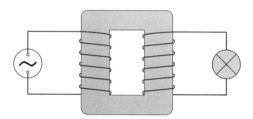

(a) Label the parts of the transformer. (4 marks)

(b) Explain how transformers are used in the National Grid system. (4 marks)

The answer is worth 5 out of the 6 marks.

The responses worth a mark are underlined in red.

We can improve the answer in several ways:

To gain the last mark the student needed to say that the movement of the cone causes **movement of the air, producing a sound wave**.

The diagram shows a moving coil loudspeaker.

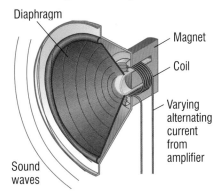

The loudspeaker contains a moveable coil attached to a cone. The cone fits loosely over a cylindrical permanent magnet. An amplifier produces a varying, alternating current in the coil.

Explain how the loudspeaker makes use of the motor effect to produce a sound wave. *(6 marks)*

A coil that carries a current and is in a magnetic field experiences a force.
The current from the amplifier varies so the size and direction of the force varies.
The force makes the coil move. This is attached to the cone, so the cone moves.

The answer is worth 3 out of the 6 marks in this Higher Tier question.

The responses worth a mark are underlined in red.

We can improve the answer in several ways:

The student leaves the answer blank, so cannot score a mark. A transformer that turns a smaller voltage into a larger voltage is called a step-up transformer.

The student uses the correct equation and rearranges it correctly, but the numbers are substituted incorrectly so the final answer is wrong. He loses the last two marks.

(a) A transformer is being used to power a 6V lamp from a 1.5V supply.
(i) What type of transformer is this? *(1 mark)*
(ii) There are 6 turns on the primary coil. How many turns are there on the secondary coil? *(3 marks)*
(b) The same transformer is used with a 3V supply.
What is the new potential difference across the secondary coil? *(2 marks)*

(a) (i) ???
(ii) $V_s/V_p = N_s/N_p$
$N_s = V_s N_p / V_p$
$N_s = 6 \times 1.5 / 6$
$N_s = 1.5$
(b) $V_s = N_s V_p / N_p$
$V_s = 1.5 \times 3V / 6$
$V_s = 0.75V$

The student has shown his working. The examiner can see that he has used the incorrect answer from part (a) correctly, so he scores both marks.

Chapter 1

Pre Test

1 Radiation.
2 Radiation.
3 White surfaces are poor absorbers of thermal radiation, so a house that has been painted white will stay cooler.
4 A light, shiny teapot will keep tea hot the longest.
5 In metals the normal conduction process occurs, but heat energy is also passed through the metal by free electrons.
6 An insulator is a poor conductor – it does not allow heat energy to pass through it easily.
7 Convection.
8 Convection is the movement of heat energy through a fluid by convection currents, caused by changes in its density due to temperature differences.
9 Convection.
10 Conduction.

Check yourself

1.1
1 Infra-red.
2 By radiation through space.
3 No, the hotter an object the more heat it radiates.

1.2
1 The best emitters of heat radiation are dark, matt surfaces.
2 The best absorbers of heat radiation are dark, matt surfaces.
3 To maximise the amount of heat they radiate.

1.3
1 Air is a poor conductor, so materials that trap air are good insulators.
2 Metal is a good conductor so the food is heated quickly, the wooden handle is a good insulator so you will not get burnt when you pick the saucepan up.
3 The best conductors of heat are metals.

1.4
1 Convection occurs in fluids.
2 The density decreases.
3 The particles in a solid are not free to move.

1.5
1 Conduction.
2 It traps air in small pockets, so it reduces convection.
3 Loft insulation is usually made from fibreglass.

End of chapter questions

1 Conduction, convection and radiation.
2 Conduction.
3 The bottom of the fluid is heated, it becomes less dense and rises and is replaced by colder, denser fluid which in turn is heated and rises.
4 If one end of the metal is heated the particles at that end gain kinetic energy and vibrate more. This energy is passed to neighbouring particles and in this way the heat is transferred through the metal. Also the free electrons in the metal gain kinetic energy and move through the metal, transferring energy by colliding with other particles.
5 Its surface area, the type of surface and its temperature compared to the surroundings.
6 Conduction involves particles vibrating and passing their energy to neighbouring particles, in a gas the particles are not held closely together.
7 It has no effect.
8 A large surface area increases the amount of radiation.
9 Concrete conducts heat away from your feet, so they feel cold. The carpet is an insulator.
10 The jacket is an insulator and reduces heat losses from the tank keeping the hot water hotter for longer.

Chapter 2

Pre Test

1 Gravitational potential energy.
2 Movement energy.
3 Sound energy to electrical energy.
4 Energy cannot be created or destroyed, only changed from one form to another.
5 Electrical energy to kinetic energy.
6 It is transferred to the surroundings, which become warmer.
7 Efficiency does not have a unit.
8 The joule, J.

Check yourself

2.1
1 Elastic potential (strain) energy.
2 From the food you eat.
3 Kinetic energy.

2.2
1 Chemical energy into heat energy.
2 Electrical energy to light and heat energy.
3 Chemical energy in the muscles, from the food you eat, is turned into kinetic energy.

2.3
1 The useful energy transformation is from electrical energy to light, energy is wasted as heat which warms the surroundings.
2 It becomes difficult to collect the spreading heat energy to turn it into other forms of energy.
3 Devices have vents so that wasted energy in the form of heat warms the surroundings. Otherwise the device would overheat.

2.4
1 The efficiency goes up.

2 Efficiency $= \dfrac{5\,J}{20\,J} = 0.25$ (25%)

3 Efficiency $= \dfrac{(5000 - 2000)}{5000} = 0.6$ (60%)

End of chapter questions

1 Gravitational potential energy – book on a shelf.
  Elastic potential energy – in a stretched bungee rope.
  Chemical energy – stored in a piece of coal.
2 Chemical energy in your muscles is changed to kinetic energy as you move, and this is changed into gravitational potential energy as you climb higher above the ground. When you fall, gravitational potential energy is changed into kinetic energy and this is changed to heat and some sound as you hit the ground.
3 In an electric heater, the useful energy change is to heat energy. As this is the same as the wasted energy, the device can be 100% efficient.

4 Efficiency $= \dfrac{720\,000\,J}{750\,000\,J} = 0.96$ (96%)

  The energy not used to heat the water heats the body of the kettle and the surroundings.
6 Elastic potential energy.
7 Electrical energy to kinetic energy.
8 Light to electrical energy.

Chapter 3

Pre Test

1 Electrical energy to heat energy.
2 Electrical energy to light and heat energy.
3 The power of a device is the rate at which it transforms energy.
4 watt, W.
5 kilowatt-hour.
6 Energy transferred = power of device × time in use
7 The National Grid is a network of transformers, pylons and cables that connects power stations to homes, schools offices and other buildings.
8 A step-up transformer.

Check yourself

3.1
1 Electrical energy to kinetic energy.
2 From chemical energy that is stored in a battery then changed to electrical energy.
3 Electrical energy to kinetic energy and thermal energy.

3.2
1 30 000
2 2.5 kW = 2500 W. So the 3000 W heater is more powerful.

3 Power $= \dfrac{36\,000\,J}{3 \times 60\,s} = 200\,W$

3.3
1 Power
2 Energy transferred = 9 kW × 20/60 h = 3 kWh
3 Energy transferred = 0.1 kW × 4 h = 0.4 kWh
  Cost = 8p × 0.4 kWh
  Cost = 3.2 p

3.4

1 A Step-down transformer.
2 High voltages reduce energy losses in the cables of the National Grid, making the system more efficient.
3 A Step-down transformer.

End of chapter questions

1 Joule per second, J/s
2 Microphone.
3 Speaker.
4 3000 J.
5 Power $= \dfrac{36\,000\,000\,J}{1 \times 60 \times 60\,s} = 10\,000\,W$
$= 10\,kW$
6 Energy transferred $= 1.2\,kW \times 10/60\,h = 0.2\,kWh$
Cost $= 0.2\,kWh \times 7p\ per\ kWh = 1.4p$
7 See diagram at top of page 13.
8 The higher voltages used in the Grid would be dangerous in the home.

Chapter 4

Pre Test

1 Coal, oil and gas.
2 Uranium/plutonium.
3 Falling water.
4 An electricity generator on top of a tall tower with blades that rotate when the wind passes over them.
5 A solar cell produces electricity from light.
6 Geothermal energy is energy released by radioactive substances deep inside the Earth, which heats the surrounding rocks.
7 Wind energy, wave energy.
8 A nuclear power station does not produce greenhouse gases.

Check yourself
4.1

1 Plutonium.
2 Nuclear fission.
3 The steam drives turbines that rotate generators and produce electricity.

4.2

1 The energy source for a hydroelectric power station is falling water, so the station must be built at the bottom of a hill for the water to fall.
2 Gravitational potential energy.
3 The tides always go in and out twice a day. Waves are generated by the wind so if there is no wind there are few waves.

4.3

1 Geothermal energy comes from radioactive processes in the rocks deep in the Earth.

2 Light energy to electrical energy.
3 A single solar cell only produces a small amount of electrical energy, even when the Sun is shining brightly, so to power larger devices many cells are joined.

4.4

1 Non-renewable.
2 Gas and hydro-electric power stations.
3 A large, flat, exposed area away from housing.

End of chapter questions

1 Renewable energy resources can be produced at the same rate as they are used up. Non-renewable energy resources are used up at a much faster rate than they can be produced.
2 Energy from falling water, waves and tides.
3 In a few parts of the world, hot water produced by geothermal energy comes up to the surface naturally and can be piped to nearby buildings to heat them.
In other places, very deep holes are drilled down to the hot rocks. Cold water is pumped down to the hot rocks where it is heated and comes back to the surface as steam. It is used to drive turbines and produce electricity.
4 Fossil fuels are a reliable energy resource. Gas-fired power stations can be started up quickly in order to cope with periods of sudden demand.
5 Fossil fuels are non-renewable so they will run out in the near future. Fossil fuels produce polluting gases such as CO_2 and SO_2.
6 Produces nuclear waste that is difficult to dispose of and there is a risk of a major accident if nuclear radiation escapes.
7 Using geothermal energy in most places requires drilling large distances through cooler rock to reach the hot rocks. Unless the hot rocks are relatively close to the surface this is too expensive.
8 Black, for better absorption of radiation.

EXAMINATION-STYLE QUESTIONS

1 (a) Light to electrical (1 mark)
  (b) (i) 2 kW = 2000 W (1 mark)
      2000 W = 2000 Joules each second (1 mark)
    (ii) Minimum time when producing maximum power (1 mark)
    time $= \dfrac{energy}{power} = \dfrac{30\ kWh}{5\ kW}$ (1 mark)
    time = 6 hours (1 mark)

(c) E.g. powering a calculator or a watch (1 mark)
2 (a) Electrical to light and sound (3 marks)
  (b) Energy is wasted as heat (1 mark)
    this would make TV hot and could be dangerous (1 mark)
    ventilation slots allow heat to heat up air and move away from TV (1 mark)
3 (a) Gravitational potential energy (1 mark)
    to kinetic energy (1 mark)
    to gravitational potential energy (1 mark)
  (b) Repeat the measurement (1 mark)
    several times and find the mean (1 mark)
4 (a) If one end of the metal is heated the particles at that end gain kinetic energy (1 mark)
    and vibrate more (1 mark)
    making neighbouring particles vibrate more (1 mark)
    so heat is transferred through the solid Metals contain free electrons (1 mark)
    these move through the metal colliding with other particles and transferring energy (1 mark)
  (b) Fluids are liquids and gases so bonds between particles are weak (1 mark)
    so difficult for neighbouring particles to pass energy on (1 mark)
5 (a) Cavity walls are filled with a material such as foam (1 mark)
    that traps air in pockets (1 mark)
    so convection currents cannot be set up in the cavity (1 mark)
  (b) Double glazing traps a layer of air between glass (1 mark)
    air is a poor conductor (1 mark)
  (c) Glass fibre traps air (1 mark)
    heat loss by conduction through roof is reduced (1 mark)
6 (a) Energy resources that are replaced as fast as they are used (1 mark)
  (b) Tidal barrages result in large areas of land being flooded (1 mark)
    destroying the habitat of birds and animals (1 mark)
    Wind farms require many large turbines which can be unsightly (1 mark)
    noisy (1 mark)
    and dangerous to birds (1 mark)

P1b   Answers to questions

Chapter 5

Pre Test

1 Radio waves.
2 Metre, m.
3 X-rays are used to take shadow pictures of bones.
4 Gamma rays are used to kill cancer cells.
5 Ultraviolet radiation can tan the skin. It can

also cause sunburn, skin ageing, skin cancer and damage to the eyes.
6 Visible light is the part of the electromagnetic spectrum that is detected by the eyes.
7 For cooking and communications.
8 Infra-red radiation.
9 A very thin glass fibre.
10 Visible light and infra-red radiation.

11 Sequence of pulses that are (usually) either on (1) or off (0).
12 The strengthening of a signal.

Check yourself
5.1

1 Hertz, Hz.
2 Gamma rays.
3 Gamma rays.

5.2
1 Gamma rays kill bacteria.
2 The lead absorbs X-rays.
3 Bones absorb X-rays but they pass through soft tissue. So the areas on an X-ray film that are *not* exposed show where the bones are.

5.3
1 By covering with clothing or suncream.
2 The markings are invisible when viewed under visible light, but show up when an ultraviolet light is shone on them.
3 Lower.

5.4
1 Infra-red cameras detect infra-red radiation given out by the survivors' bodies which are warmer than the surroundings.
2 Microwaves are strongly absorbed by the water molecules inside food, heating them.
3 Radio waves are produced by applying an alternating voltage to an aerial.

5.5
1 Visible light and infra-red radiation.
2 By continuous reflections along the fibre.
3 Microwaves.

5.6
1 Signals weaken over distance.
2 Noise picked up by the signal is amplified along with the original signal.
3 Digital

End of chapter questions
1 Electromagnetic waves are electric and magnetic disturbances that travel as waves and move energy from place to place.
2 They all travel at the same speed in a vacuum.
3 Radio waves.
4 Sterilising surgical instruments, keeping food fresh by killing bacteria on it and killing cancer cells.
5 TV and video remote controls and cooking food.
6 A signal that varies continuously in amplitude.
7 They wear lead aprons and stand behind lead screens when taking X-rays.
8 The wavelength is shorter.
9 Microwaves are strongly absorbed by the water in the cells, causing heating and damage.
10 The frequency is lower.
11 Optical fibres are passed into the body and light is transmitted along them so that images of the inside of the body can be seen.
12 Speed = frequency × wavelength

Chapter 6

Pre Test
1 A small central nucleus made up of protons and neutrons, surrounded by electrons.
2 It has no effect.
3 An alpha particle consists of 2 protons and 2 neutrons.
4 A few centimetres.
5 It decreases.
6 It decreases to half the original value.
7 In smoke alarms.
8 Alpha particles.

Check yourself
6.1
1 The nucleus.
2 Isotopes of an element have the same number of protons, but different numbers of neutrons, in the nucleus.
3 Nothing, the rate of decay is independent of the temperature.

6.2
1 Gamma rays.
2 Alpha particles.
3 Gamma radiation is an electromagnetic wave from the nucleus.

6.3
1 It decreases.
2 It has decreased to one quarter of its original value.
3 It has decreased to one quarter of its original value.

6.4
1 It is less ionising than alpha radiation.
2 The half life must be long enough to complete the medical procedure, but short enough to avoid exposing the patient for longer than is necessary.
3 Alpha particles are very poorly penetrating so they would not be detected outside the body.

End of chapter questions
1 A beta particle is an electron from the nucleus.
2 A few metres.
3 The time it takes for the number of radioactive atoms in a sample to halve.
4 Gamma radiation is a high frequency electromagnetic wave emitted from the nucleus of an atom.
5 Gamma rays.
6 It drops to 1/8 of the original value.
7 Gamma rays.
8 Beta particles.

Chapter 7

Pre Test
1 The sound waves you hear would be a different frequency (pitch) to the sound waves emitted by the source.
2 The Milky Way.
3 Its frequency changes.
4 With a massive explosion at a very small initial point – the Big Bang.
5 A device that collects light or other electromagnetic radiations so distant objects can be observed.
6 Telescopes are used on the Earth or on satellites in space.

Check yourself
7.1
1 A galaxy is a collection of millions of stars.
2 Red shift is the changing of the frequency of light from distant galaxies towards the red end of the spectrum, because they are moving away from us.
3 The most distant ones.

7.2
1 It shows us that distant galaxies are moving away from us, and the furthest ones are moving the fastest, so the Universe is expanding outwards in all directions.

2 The massive explosion that was the start of the Universe.
3 If the Universe is now expanding outwards in all directions, this suggests that it started with an explosion from a single point.

7.3
1 The layer of gas that surrounds the Earth.
2 It absorbs and scatters it.
3 The signals received by telescopes on satellites are not distorted by the atmosphere.

End of chapter questions
1 Billions.
2 A blue shift would show that they are moving towards us.
3 Distant galaxies are moving away from us, the most distant galaxies are moving the fastest and this is true of galaxies no matter which direction you look.
4 The frequency has decreased.
5 The images produced are clearer.
6 A telescope that collects radio waves rather than visible light.

EXAMINATION-STYLE QUESTIONS

1 (a) (i) gamma radiation   (1 mark)
       (ii) microwaves   (1 mark)
    (b) (i) tanning   (1 mark)
            burning   (1 mark)
            ageing   (1 mark)
            skin cancer   (1 mark)
       (ii) ultraviolet is absorbed by the ink
            the energy is given out as visible light   (1 mark)
    (c) they travel as waves and move energy from place to place   (1 mark)
        all electromagnetic waves travel through space at the same speed
                                       (1 mark)

2 (a) 120 ÷ 2 = 60
        60 ÷ 2 = 30
        30 ÷ 2 = 15   (1 mark)
        3 half-lives   (1 mark)
    (b) half-life = $\dfrac{180 \text{ seconds}}{3}$ (1 mark)
        half-life = 60 seconds   (1 mark)
    (c) measure activity in counts per second
        e.g. every minute   (1 mark)
        for 10 minutes   (1 mark)
    (d) graph uses line drawn from a number of measurements / produces an averaging effect   (1 mark)

3 (a) medical equipment is passed through a beam of gamma rays   (1 mark)
        killing any bacteria   (1 mark)
        so the equipment is sterilized   (1 mark)
    (b) a narrow beam of gamma rays is directed at the tumour   (1 mark)
        from different directions   (1 mark)
        to kill the tumour not the surrounding tissue   (1 mark)
    (c) X rays pass through soft tissue but are absorbed by bone   (1 mark)
        They are detected with photographic film   (1 mark)
        producing shadows where there is bone   (1 mark)

4 (a) alpha and beta   (1 mark)
        they are charged particles   (1 mark)
        gamma radiation has no charge (1 mark)
   (b) (i)   nuclear radiation collides with the
              atoms of a material   (1 mark)
              and knocks electrons off them
              creating ions   (1 mark)
        (ii) alpha is biggest particle/has
              biggest charge   (1 mark)
              so undergoes most collisions
                                           (1 mark)

        so alpha most ionising   (1 mark)
5 (a) (i)   Telescopes collect more light than
              the eye so fainter objects can be
              seen   (1 mark)
        (ii) telescope that detects radio waves
                                           (1 mark)
        (iii) wavelength of radio waves is
              longer than wavelength of light
              waves   (1 mark)
        (iv) optical telescope affected by
              amount of light   (1 mark)

        radio telescope not affected by
              cloud cover etc   (1 mark)
   (b) gamma rays not able to get through the
        atmosphere   (1 mark)
        so not detected on Earth   (1 mark)

P2 Answers to questions

Chapter 1

Pre Test
1 Metres per second, m/s.
2 Horizontal line.
3 Acceleration = $\dfrac{\text{change in velocity}}{\text{time taken for the change}}$
4 A deceleration.
5 Acceleration.
6 Horizontal line.
7 Slope = $\dfrac{\text{change in } y \text{ value}}{\text{change in } x \text{ value}}$
8 Slope increases.

Check yourself
1.1
1 Speed.
2 Speed = $\dfrac{\text{distance travelled}}{\text{time taken}}$
3 Metre, m.
1.2
1 Velocity is speed in a given direction.
2 Metres per second squared, m/s^2.
3 Slowing down.
1.3
1 A body travelling at a constant velocity.
2 Straight line graph with a negative slope.
3 Distance travelled.
1.4
1 10 m/s.
2 20 m/s.
3 1 m/s^2.

End of chapter questions
1 A body that is stationary.
2 6.25 m/s.
3 If it is changing direction, its velocity is
   changing so it is accelerating, even if it is
   travelling at a steady speed.
4 Acceleration.
5 Deceleration.
6 Area under the graph.
7 10 m/s.
8 99 m.

Chapter 2

Pre Test
1 They are equal.
2 Upwards.
3 A single force that has the same effect on a
   body as all the original forces acting together.
4 It accelerates the object in the direction of
   the force.

5 Resultant force = mass × acceleration.
6 3 m/s^2.
7 Time between the driver seeing something
   and responding to it.
8 Poor roads, poor weather, worn tyres, worn
   brakes.
9 Velocity reached by an object falling
   through a fluid when the weight is equal to
   the resistive forces.
10 Weight = mass × gravitational field strength.

Check yourself
2.1
1 N.
2 Downwards.
3 15 N to the left.
2.2
1 Continues moving at a steady speed in a
   straight line.
2 Newton.
3 When the direction of the force is opposite
   to the direction the body is moving in.
2.3
1 Mass = $\dfrac{\text{force}}{\text{acceleration}}$
2 Increases.
3 0.004 m/s^2.
2.4
1 Zero.
2 Stopping distance = thinking distance +
   braking distance.
3 The greater the speed, the greater the
   stopping distance.
2.5
1 N/kg.
2 Initially the only force acting is the weight
   downwards.
3 It reaches terminal velocity.

End of chapter questions
1 5 N.
2 They are opposite.
3 Accelerates in the direction of the force.
4 It decelerates.
5 It accelerates in the direction of the resultant
   force.
6 140 N.
7 The distance travelled between the driver
   seeing something and reacting to it by
   putting on the brakes.
8 Driver tired or under the influence of alcohol
   or drugs.
9 Weight.
10 Zero.

Chapter 3

Pre Test
1 Energy transferred.
2 Work done = force × distance moved in the
   direction of the force.
3 Kinetic energy.
4 Energy stored in an elastic object when it is
   stretched or squashed.
5 Momentum = mass × velocity.
6 Always applies when bodies interact,
   provided no external forces act.
7 0.5 m/s.
8 0.3 m/s.
9 Crumple zone increases the time for the
   momentum to become zero, so force on the
   car decreases.
10 16 000 kg m/s.

Check yourself
3.1
1 Joule.
2 When it moves through a distance.
3 2100 J.
3.2
1 Kinetic energy increases.
2 Joule.
3 50 000 J.
3.3
1 When they are moving.
2 Kg m/s or Ns.
3 10 000 kg m/s.
3.4
1 Zero.
2 They will be equal and opposite.
3 The student with twice the mass has half the
   speed.
3.5
1 Change in momentum = force × time taken
   for the change.
2 2500 N.
3 The change in momentum is the same
   whatever she lands on. On the mat the time
   taken for the change is increased so the
   force on her is decreased.

End of chapter questions
1 Joule.
2 400 N.
3 Kinetic energy = $\frac{1}{2}$ mass × speed^2.
4 80 kg.
5 40 000 kg m/s.
6 0.6 m/s.
7 Total momentum before the collision.
8 Velocity has direction.

9 2.5 N.
10 16 000 kg m/s.

EXAMINATION-STYLE QUESTIONS

1 (a) (i) Weight.   (1 mark)
       (ii) Drag or air resistance or air friction.
                                                         (1 mark)
   (b) Flying at a constant forward speed,
                                                         (1 mark)
        and a constant height.   (1 mark)
   (c) (i) Lift.   (1 mark)
        (ii) Thrust.   (1 mark)
2 (a) weight = 50 kg × 10 N/kg   (1 mark)
        weight = 500 N   (1 mark)
   (b) work done = 200 N × 4.0 m   (1 mark)
        work done = 800 J   (1 mark)
3 (a) (i) change in momentum =
             force × time   (1 mark)
        (ii) change in momentum =
             200 N × 0.02 s
             change in momentum = 4 Ns
                                                         (1 mark)
             change in momentum =
             mass × change in velocity   (1 mark)
             velocity = 4 Ns/0.16 kg   (1 mark)
             velocity = 25 m/s   (1 mark)
   (b) kinetic energy = $\frac{1}{2}mv^2$ (1 mark)
        kinetic energy = $\frac{1}{2} \times 0.16$ kg × 25^2
                                                         (1 mark)
        kinetic energy = 50 J   (1 mark)
4 (a) Total momentum before event equals
        total momentum after event   (1 mark)
        (change in momentum is zero scores
        2 marks)
   (b) collisions   (1 mark)
        explosions   (1 mark)
   (c) momentum before = 0.2 kg × 5 m/s +
        0.3 kg × 2 m/s   (1 mark)
        momentum before = 1.6 kg m/s
                                                         (1 mark)
        momentum after = 0.5 kg m/s × v m/s
                                                         (1 mark)
        v = 1.6 kg m/s / 0.5 kg
        v = 3.2 m/s   (1 mark)
5 (a) Thinking distance is proportional to
        speed.   (1 mark)
        Braking distance increases more
        quickly as speed increases.   (1 mark)
   (b) stopping distance = thinking distance
        + braking distance   (1 mark)
        thinking distance = 17 m   (1 mark)
        braking distance = 50 m   (1 mark)
        stopping distance = 67 m   (1 mark)
   (c) (i) thinking distance same   (1 mark)
             braking distance increases   (1 mark)
        (ii) thinking distance increases   (1 mark)
             braking distance same   (1 mark)

Chapter 4

Pre Test
1 Electrons are transferred onto it.
2 They repel each other.
3 Charge moves through it easily.
4 They have free electrons.
5 They pass through a charged grid.
6 They stick to plates, on the walls, with the
   opposite charge.

Check yourself
4.1
1 Negative.

2 Electrons are rubbed off it.
3 Repulsive.
4.2
1 They do not have free charge carriers.
2 Isolate from the earth.
3 Electric current is the rate of flow of charge.
4.3
1 It prevents static charge building up on the
   pipe.
2 The paint droplets become positively
   charged as they leave the sprayer so they
   repel each other and spread out to form a
   cloud.
3 By friction with the walls of the pipe.

End of chapter questions
1 The positive charges are on the proton in
   the nucleus.
2 Friction against your body charges the
   clothing.
3 Copper or another metal.
4 Connect it to earth .
5 Electrostatic paint sprayer, photocopier,
   electrostatic smoke precipitator.
6 You would probably get a shock as the
   charge flowed through you to the earth.

Chapter 5

Pre Test
1 Two or more cells.
2
3 Potential difference = current × resistance.
4 Ampere, A.
5
6 Resistance decreases.
7 Current = $\dfrac{\text{potential difference of the supply}}{\text{total resistance of the circuit}}$
8 Charge has no choice of route, it must go
   through every component.
9 Each component is connected across the
   supply p.d.
10 Add the currents in the individual branches
    of the circuit.

Check yourself
5.1
1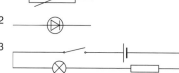
2
3
5.2
1 Ammeter.
2 Ohm, Ω.
3 Opposition to current flow.
5.3
1 Resistance increases.
2 No effect.
3 There is current in one direction, in the other
   the resistance is very high and the current is
   zero.
5.4
1 Charge stops flowing through any
   component.
2 Add the individual resistances.
3 p.d. increases.
5.5
1 Little effect on the other components.

2 All the same.
3 Add the currents in the individual branches.

End of chapter questions
1 Diagram using symbols to show
   components in a circuit.
2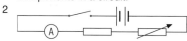
3 In series.
4 Voltmeter.
5
6 As current increases the temperature of the
   filament increases and its resistance
   increases.
7 18 Ω.
8 0.5 A.
9 p.d. stays the same.
10 Current decreases.

Chapter 6

Pre Test
1 Current that continuously changes
   direction.
2 50 Hz.
3 Brown.
4 Plastic or rubber.
5 A thin piece of wire.
6 The case will become live.
7 Power = energy/time.
8 Watt, W.
9 Charge = current × time.
10 Electrical energy to heat.

Check yourself
6.1
1 Current in one direction only.
2 Zero.
3 230 V.
6.2
1 To prevent the case giving someone a
   shock should it become live.
2 Blue.
3 Brass is a good conductor, does not
   corrode and is hard.
6.3
1 An electromagnetic switch that will
   disconnect the supply if the current exceeds
   a certain value.
2 A large current will flow to earth and melt
   the fuse.
3 Plastic is an insulator, so current cannot
   flow through to give a shock to anyone
   touching the case.
6.4
1 2300 W.
2 90 000 J.
3 3 amp fuse.
6.5
1 Coulomb, C.
2 46 000 J.
3 240 C.

End of chapter questions
1 The direction of the current reverses 50
   times each second.
2 d.c. flows in one direction only, a.c.
   continuously changes direction.
3 Tightly over the outer plastic casing.
4 Green and yellow.
5 Live.

6 It must be higher so the fuse does not melt during normal use. It should be only slightly higher or the current may become large enough to damage the appliance before the fuse melts.

7 Power = current × potential difference.

8 10 A.

9 Energy transformed = potential difference × current.

10 Charge.

Chapter 7

Pre Test

1 The number of protons in the nucleus.

2 The number of protons and neutrons in the nucleus.

3 The atom is a blob of positive matter with negative charges stuck in it.

4 Thin gold foil.

5 The splitting of a nucleus.

6 There are 238 protons and neutrons in the nucleus.

7 The joining together of two nuclei.

8 They all have a positive charge and like charges repel.

Check yourself
7.1

1 Cosmic rays from space, rocks, nuclear power stations, etc.

2 It stays the same.

3 Goes down by two.

7.2

1 Alpha particles were fired at a piece of thin gold foil.

2 Atoms consist mostly of empty space.

3 It is where the mass of the atom is concentrated and is positively charged.

7.3

1 Uranium containing a higher percentage of uranium 235 than occurs naturally.

2 A uranium or plutonium nucleus absorbs a neutron.

3 Uranium 235 and plutonium 239.

7.4

1 Fusion.

2 They must be raised to a very high temperature.

3 By a magnetic field.

End of chapter questions

1 Mass number goes down by four.

2 Atomic number goes up by one.

3 Alpha particles.

4 Empty space.

5 An isotope that will undergo fission if it absorbs a neutron.

6 When a fission occurs, two or three neutrons are produced. If these go on to cause fissions that produce more neutrons that go on to produce more fissions, a chain reaction occurs that gets bigger and bigger.

7 In stars.

8 All nuclei have a positive charge so they repel each other.

EXAMINATION-STYLE QUESTIONS

1 (a) (i) 1.5 V   (1 mark)

      (ii) 1.5 V   (1 mark)

   (b) (i) current = p.d./resistance   (1 mark)

        current = 1.5 V/3 Ω

        current = 0.5 A   (1 mark)

      (ii) current = 1.5 V/6 Ω

        current = 0.25 A   (1 mark)

   (c) current = 0.5 A + 0.25 A   (1 mark)

       current = 0.75 A   (1 mark)

2 (a) current = power/potential difference   (1 mark)

        current = 2300/230 V   (1 mark)

        current = 10 A   (1 mark)

   (b) 13 A   (1 mark)

   (c) time = charge/current   (1 mark)

        time = 2400 C/10 A   (1 mark)

        time = 240 s   (1 mark)

        time = 4 minutes   (1 mark)

3 (a) Y and Z.   (1 mark)

   (b) X and Y.   (1 mark)

   (c) X and Y.   (1 mark)

   (d) It must lose an electron.   (1 mark)

   (e) Beta decay – 1 more proton,   (1 mark)

       1 less neutron.   (1 mark)

4 (a) correct symbols for cell, resistor and variable resistor   (3 marks)

       all connected in series   (1 mark)

   (b) (i) in series   (1 mark)

      (ii) in parallel   (1 mark)

   (c) Straight line with positive slope (1 mark)

       passing through the origin   (1 mark)

5 Neutron is absorbed by a large nucleus   (1 mark)

Large nucleus splits into 2 or 3 smaller nuclei   (1 mark)

Plus more neutrons   (1 mark)

Releasing energy   (1 mark)

Neutrons go on to be absorbed by other nuclei   (1 mark)

This is a chain reaction   (1 mark)

P3   Answers to questions

Chapter 1

Pre Test

1 A moment.

2 Apply the force further from the pivot.

3 The point on an object where all the mass of the object can be thought to be concentrated.

4 At the centre.

5 The principle that says the clockwise moment is equal to the anticlockwise moment.

6 When the object is in equilibrium.

7 The size of its base and the position of its centre of mass.

8 When its centre of mass moves outside its base.

9 Acceleration towards the centre of a circle.

10 Towards the centre of the circle.

11 The attractive force between masses.

12 It gets less.

13 Slightly squashed circles (ellipses).

14 Gravitational force.

15 A satellite used to send and receive telephone and TV signals.

16 A geostationary orbit.

Check yourself
1.1

1 12.0 N m

2 40 N

3 A crowbar allows you to apply the same force at a bigger distance from the pivot, giving a bigger moment.

1.2

1 (a)

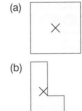

(b)

(c)

1.3

1 1 m.

2 450 N.

3 The perpendicular distance to the pivot is zero.

1.4

1 The bags raise the centre of mass so it will not have to tilt so far before the line of action of the weight moves outside the base.

2 It means that they are likely to topple over when hit by a bowling ball.

3 C.

1.5

1 It is constantly changing direction, so constantly changing its velocity, so accelerating.

2 The tension in the string.

3 It will fly off on a path at a tangent to the circle at the point where the string breaks.

1.6

1 The force increases.

2 The force of gravity acting on you would be less.

3 The gravitational force is very small because the masses are relatively small.

1.7

1 It is shorter than the time taken for any of the other planets to complete one orbit.

2 Mercury moves more quickly in its orbits than any of the other planets.

3 It is the gravitational force that provides the centripetal force to keep Mercury in orbit.

1.8

1 24 hours.

2 A particular satellite provides coverage for a particular region on the surface of the Earth. If it moved, the coverage would no longer be provided.

3 The Moon.

End of chapter questions

1 The turning effect of a force.

2 The size of the force and the perpendicular distance from the line of action of the force to the pivot.

3 Vertically below the point of suspension.

4 Over the knife-edge.

5 They are equal.

6 50 cm from the centre on the other side.

7 To make it more stable.

8 The weight produces a moment that makes the object rotate and topple.

9 Towards the centre of the circle.

10 The mass of the body, the speed of the body and the distance from the centre of the circle.

11 All objects with mass.

12 Their masses and the distance between their centres.

13 The path of one body around another.

14 The time taken increases as the distance from the Sun increases.

15 A satellite designed to pass over the whole Earth and observe it.

16 A polar orbit.

Chapter 2

Pre Test

1 The angle between an incident ray and the normal.

2 The angle between a reflected ray and the normal.

3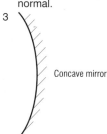

Concave mirror

4 The point where parallel rays of light are brought to focus by the mirror.

5 The change in direction of light as it passes the boundary from one material into another.

6 It gets less.

7 Thicker in the middle than the ends.

8 A virtual image.

9 A converging lens.

10 A converging lens.

11 An echo.

12 Sound waves cannot travel through a vacuum.

13 The frequency of the sound wave.

14 It becomes louder.

15 Sound with a frequency above 20 000 Hz.

16 Ultrasound is used to make images of soft structures, such as an unborn baby.

Check yourself

2.1

1 The same size as the object.

2 1.5 m behind the mirror.

3 A virtual image.

2.2

1 1.4.

2 Convex mirrors.

3 The point where parallel rays of light are brought to focus by the mirror.

2.3

1 Dispersion.

2 Red.

3 Light waves change speed and direction as they travel from one material into another.

2.4

1 Between the principal focus and the lens.

2 The distance from the centre of the lens to the principal focus.

3 Upright.

2.5

1 Rays drawn on a ray diagram so that the position of the image can be found.

2 A real image.

3 A diverging lens does not produce a real image that can be formed on a film.

2.6

1 One where the direction of vibration of the particles is in the same direction as the direction of travel of the waves.

2 One where the direction of vibration of the particles is at right angles to the direction of travel of the waves.

3 The range varies with such things as age, but is approximately 20 Hz to 20 000 Hz.

2.7

1 They have different waveforms.

2 The frequency increases.

3 It causes mechanical vibrations in the air around it.

2.8

1 Flaws in the metal act as boundaries and reflect the ultrasound back.

2 Ultrasound is non-ionising, so it will not harm the baby.

3 Ultrasound can be used to clean the mechanism without having to take it apart, reducing the risk of damage.

End of chapter questions

1 A line drawn at 90° to a reflecting or refracting surface at the point where an incident ray meets the surface.

2 The angle of incidence is equal to the angle of reflection.

3

Convex mirror

4 A concave mirror can form a real image or a virtual image depending on the position of the object.

5 It speeds up.

6 It bends towards the normal.

7 It is thinner in the middle and thicker at the ends.

8 The point where rays of light parallel to the principal axis are brought to focus.

9 It can produce a real image that is smaller than the object.

10 A virtual, magnified image.

11 Sound waves are mechanical vibrations in a substance, so they need a substance in which to travel.

12 Longitudinal waves.

13 The amplitude of the wave.

14 The pitch decreases.

15 The frequency is too high to be detected by the human ear.

16 It is safer than X-rays.

EXAMINATION-STYLE QUESTIONS

1 (a) moment = force × perpendicular distance

moment = 420 N × 2 m   (1 mark)

moment = 840 N m   (1 mark)

(b) 840 N m = weight × 1.6 m   (1 mark)

weight = 840 N m / 1.6 m

weight = 525 N   (1 mark)

(c) Derek's weight is bigger so his moment is bigger.   (1 mark)

Derek's side of the seesaw goes down / Laurie's side goes up.   (1 mark)

2 (a) The point at which its mass (seems to) act.   (1 mark)

(b) Any five from: clamp pin horizontally; hang card from pin through hole; hang plumbline from pin; mark position of plumbline on sheet; use ruler to join marks with straight line; repeat with other hole; centre of mass is where lines cross; check by balancing/repeat with third hole. [max. 3 marks if no 'repeat with other hole']   (5 marks)

3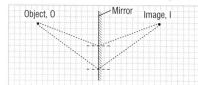

(a) 2 normals correctly drawn.   (1 mark)

2 reflected rays correctly drawn.   (1 mark)

Position of image correct.   (1 mark)

(b) One that light rays only appear to have come from.   (1 mark)

4 (a) (i) Rays of light need to cross to form an image.   (1 mark)

to affect chemical on a film etc.   (1 mark)

(ii) Object is beyond the principal focus.   (1 mark)

(b) (i) Object between the principal focus and lens.   (1 mark)

(ii) Magnification = image height / object height   (1 mark)

Magnification = 4.0 cm / 1.6 cm

Magnification = 2.5   (1 mark)

5 (a) Sound with frequency above audible range (>20 000 Hz).   (1 mark)

(b) Object suspended in fluid.   (1 mark)

Ultrasound generator makes fluid particles vibrate. (1 mark)
Fluid particles knock small dirt particles off watch. (1 mark)
(c) Ultrasound waves sent into metal object. (1 mark)
Receiver picks up reflected waves. (1 mark)
Reflections shown on oscilloscope. (1 mark)
Flaws in object cause reflections and so are detected. (1 mark)

Chapter 3

Pre Test
1 It experiences a force (unless it is parallel to the magnetic field lines).
2 The size of the force increases.
3 A potential difference is induced across its ends.
4 A current will flow in the circuit.
5 Two insulated coils wound on an iron core.
6 An a.c. supply.
7 A step-up transformer increases the potential difference.
8 A step-down transformer.

Check yourself
3.1
1 It stays the same.
2 It reverses the direction of the current around the coil every half-turn.
3 The force is zero.

3.2
1 The p.d. is zero.
2 A p.d. is induced, opposite in polarity to the original p.d.
3 A p.d. is induced.

3.3
1 To prevent current shorting across the iron core or across adjacent turns of wire.
2 Iron can be magnetised.
3 The transformer will not work, as it requires an a.c. supply. A cell supplies d.c.

3.4
1 So that the electrical energy can be transmitted at a high p.d., reducing energy losses in the cables.
2 8 V
3 2400.

End of chapter questions
1 A current carrying a wire in a magnetic field experiences a force; this is called the motor effect.

2 The force is perpendicular to the wire and to the magnetic field.
3 A potential difference is induced across the ends of the wire.
4 The size of the potential difference is increased.
5 Iron.
6 It sets up an expanding and collapsing magnetic field in the core of the transformer.
7 It reduces the potential difference across the primary coil to a smaller potential difference across the secondary coil.
8 At sub-stations.

Chapter 4

Pre Test
1 The Milky Way.
2 At least a billion.
3 An area with a gravitational field so intense that not even light can escape from it.
4 Nuclear fusion.
5 A process in which lighter nuclei fuse together to form heavier nuclei.
6 Hydrogen.

Check yourself
4.1
1 A group of millions of stars.
2 It decreased.
3 Millions of years.

4.2
1 An extremely dense core left after a supernova explosion.
2 No light escapes from it.
3 The force of gravity inwards is balanced by the radiation pressure outwards.

4.3
1 Nuclear fusion.
2 In the last stages of the life of a very big star before it explodes as a supernova.
3 They contain elements heavier than iron.

End of chapter questions
1 Gravitational force.
2 Dust and gas.
3 Its mass.
4 The explosion of a star much larger than the Sun near the end of its life.
5 By nuclear fusion processes in stars.
6 By supernovae explosions.

1 (a) When enough dust and gas (1 mark) are pulled together by gravitational attraction. (1 mark)
  (b) (i) Gravitational forces (1 mark) inwards, (1 mark) balanced by radiation pressure/force (1 mark) outwards. (1 mark)
    (ii) Swells to become a red giant, (1 mark) cools and contracts, (1 mark) becomes a white dwarf, (1 mark) fades to a black dwarf. (1 mark)

2 (a) Fusion (1 mark)
    heavier (1 mark)
    energy (1 mark)
    supernova. (1 mark)
  (b) A point where gravitational field is so strong that (1 mark) not even light can escape from it. (1 mark)

3 (a) [0] (1 mark)
  (b) [0] (1 mark)
  (c) [0] (1 mark)
  (d) [0] (1 mark)
  (e) [0] (1 mark)

4 (a)

Iron core
Primary coil   Secondary coil
a.c. generator   Lamp
(4 marks)

  (b) Step-up transformers increase p.d. from power stations, (1 mark) to reduce energy losses in transmission. (1 mark) Step-down transformers reduce p.d. in sub-stations, (1 mark) to reduce p.d. to safe level for consumers. (1 mark)